Paths of Destiny

Paths of
Destiny

E. V. Thompson

timewarner books

A *Time Warner* Book

First published in Great Britain in 2003
by Time Warner Books

Copyright © 2003 by E. V. Thompson

The moral right of the author has been asserted.

A CIP catalogue record for this book
is available from the British Library.

HARDBACK ISBN 0 316 85721 1
C FORMAT ISBN 0 316 85723 8

Typeset by Palimpsest Book Production Limited,
Polmont, Stirlingshire
Printed and bound in Great Britain by
Clays Ltd, St Ives plc

Time Warner Books UK
Brettenham House
Lancaster Place
London WC2E 7EN

www.TimeWarnerBooks.co.uk

Paths of Destiny

1

I

Descending the servants' stairs from her attic room in Treleggan rectory at 7 a.m. on a warm August morning, Alice Rowe made her way along the passageway to the kitchen with an uneasy premonition that all was not as it should be.

She could not account for the feeling. At least, not immediately. Everything seemed exactly as it had been when she finished work the previous evening and went upstairs to her room.

Looking about her when she reached the kitchen, she suddenly realised it was this very fact that was troubling her. Everything was *too* tidy. Nothing had been disturbed during her absence from the main living quarters of the large house.

When she checked the front door, she discovered it was still unbolted. She had left it like that last night because Parson Markham had not returned from his daily round of visiting parishioners.

Finding it unlocked did not worry her unduly. The parson was well into his seventies and increasingly forgetful. Besides, Treleggan was one of the most remote villages on Bodmin Moor and unlikely to attract the attention of casual burglars.

She decided she would go about her work for a while and stop worrying, but she did not find it easy. Her concern was for the kindly old parson, but she was also aware that if anything happened to prevent him continuing his duties at Treleggan, it was unlikely she would be able to remain in the village.

An illegitimate child, Alice had spent the first six months of her life in Bodmin gaol because of her mother's refusal to give the name of Alice's father to the magistrate.

There had been a reason for Grace Rowe's silence, but it was not one she would disclose to the court. Her lover was a miner who, three months before Alice's birth, had been seriously injured in an accident at work. She would not add to his problems by having a bastardy order served upon him.

Not until the miner died did Alice's mother name him and so secure her release from prison. But her troubles were far from being at an end. Treleggan was a tight-knit farming community, in constant and bitter dispute with the moorland tin miners.

Grace Rowe's moral lapse would eventually have been accepted had it occurred with a village man. The fact that her lover was a miner was unforgivable. The village shunned her and she was forced away.

For a number of years mother and daughter suffered a precarious and nomadic existence in Cornwall until the life they were leading led to a serious deterioration in Grace's health. Unable to work, she was committed to a workhouse, the law decreeing it should be in the area in which she was born. Mother and daughter were sent to Liskeard, a town no more than five miles from Treleggan.

Here they were discovered by the Reverend Arnold Markham, during one of the visits he was in the habit of making to the workhouse.

As Grace Rowe's health deteriorated still further she realised she was dying and begged the Treleggan parson to help her daughter find work and so escape from life as a pauper.

When Alice's mother died, the parson made good his promise, taking Alice into the parsonage as a housemaid to help out his ageing housekeeper.

The Treleggan villagers were outraged. Not only had he taken a thirteen-year-old girl into his bachelor establishment – but she was the bastard daughter of a miner!

For a time the villagers shunned the parson, his church and Alice. Although most eventually returned to the fold, they never accepted Alice as one of their own.

Now twenty years of age, Alice had been unable to make friends in the village. Had it not been for the extreme kindness of her employer, she would have led an unhappy and lonely life. As it was, she had come to look upon Parson Markham as a father figure and she adored him.

He, in his turn, looked upon her as a daughter and had delighted in giving her the schooling she would never have received had she been brought up as a normal village girl.

In recent years, realisation had come to Alice that such a life could not last for ever. Parson Markham was an old man and increasingly frail. Should he be forced to retire, a new rector would be appointed to Treleggan. The new incumbent would gain immediate favour with his parishioners by dismissing her.

These were the thoughts that passed through Alice's mind as she went about her work that August morning. Eventually, unable to concentrate on anything for very long, she knew she had to check on the well-being of her employer.

There was an additional reason for her unease. She had heard no sound from Digger, the Jack Russell terrier which slept in a basket in Parson Markham's bedroom.

Twelve years old, the small dog was not always aware of Alice descending the rear staircase, yet, no matter how silent she tried to be, it would hear her when she began moving about in the kitchen. Whining and scratching at the bedroom door, it would not rest until the elderly rector rose from his bed to put out his pet, grumbling about having his routine dictated by a diminutive dog.

Once the bedroom door was opened, Digger would scoot at speed along the passageway to the stairs. By the time Parson Markham had climbed back into bed, still grumbling, Digger would be in the kitchen. Dancing excitedly around Alice, his short stump of a tail twitching with metronomic frenzy, the lively little dog knew that only a brief visit to the rectory garden stood in the way of the first meal of the day.

But today there had been no sound from the bedroom and Digger had not put in an appearance.

Climbing the stairs slowly, Alice hoped in vain that Digger might appear before she reached the first floor.

At the door of the parson's bedroom she paused for a moment, listening for a sound from within. Hearing nothing, she knocked at the door, nervously at first, but louder when there was no response.

Apprehensively, Alice opened the door, fearful of what she might find inside. It came as a momentary relief to discover the bedroom was empty.

At first, she thought Parson Markham might have been called out unexpectedly earlier that morning, perhaps to pray with a dying parishioner. She would not have heard the doorbell from her attic room. However, when she saw his bed had not been slept in her misgivings returned, and now she was seriously concerned. On the rare occasions when the parson intended being absent from the rectory overnight he would tell her in advance, so she could bolt the door when her work was done.

Alone in the house until Ivy Deeble, the rector's housekeeper, arrived from the home she shared with her sister in the village, Alice was momentarily at a loss about what she should do. Then she remembered the diary in which the parson kept details of his appointments and the visits he proposed making to his parishioners. They were always scrupulously recorded. The Reverend Markham was fully aware his memory was not as accurate as it had once been.

The leather-bound diary was kept on the desk in the rectory study. Opening it, Alice riffled through the pages until she found the one for Thursday, 10 August, 1854.

There were only two entries for the day. Both were visits to parishioners who lived outside the village. The first was to Tor Farm, a lonely moorland holding worked by an eccentric bachelor.

4

The second was to the cottage of the redoubtable Widow Tabitha Hodge. Now ninety years of age, Widow Hodge still ran the disreputable 'kiddleywink' – drinking-house – she had opened in the cottage built by her miner husband in the year of their marriage, seventy years before.

Always well stocked with smuggled spirits, the kiddleywink was popular with the tough and often unruly miners who worked the tin mines on the high moor. Attempts by the rector and others to persuade the nonagenarian widow to move to a less remote dwelling were met with the determined proclamation that she would remain in her matrimonial home 'until the Lord took her to join her late husband'.

Alice decided she would make a visit to the widow her first call.

II

Stopping at the cottage that Ivy Deeble shared with her sister, Alice left a message to be passed on to the almost totally deaf housekeeper, who was still in bed. Then, leaving the village behind, she struck out across the open moor, heading for the isolated kiddleywink.

She found Widow Hodge standing in a small, grassless yard between cottage and barn, scattering corn for a couple of dozen chickens, a few ducks and two enormous turkeys.

Born into a Church of England family, Tabitha Hodge might have been forgiven had she converted to Wesleyism. As a small girl, she had been held on the lap of John Wesley, who praised her as 'the prettiest little girl he had met with on his tours of the West Country'. Instead, she had shown indifference to all religious movements. She had set foot inside a church only once since her wedding day – and that was on the occasion of her husband's burial.

In spite of this and whatever the weather, Parson Markham made the journey to her cottage at least once a week to check on her well-being. He had even been known to spend an hour or so on his knees, weeding her vegetable patch.

If the old woman appreciated the rector's concern for her, she was at pains not to show such feelings to anyone. In response to Alice's question, she replied, 'Parson Markham? I haven't seen him for a week or more, and if he's got any sense he'll stay away. I've told him more than once that he's getting too old to come tramping around the moor in all weathers.'

Although Arnold Markham had been rector of Treleggan for more than thirty years, he had arrived there from an urban parish. Alluding to this, Widow Hodge opined, 'It don't do for them as don't know the moor to wander too far from home.' Mischievously, she added, 'Perhaps the same fox that made off with two of my ducks in the night has taken him off too.'

The little joke amused the old woman and she was still chuckling as Alice walked away. However, Alice had not gone far before Widow Hodge called after her.

'It might be worth your while taking a look in the woods around Treveddoe. One of the miners from that way came in last night with a face white as a corpse. Said he'd heard such a howling up there he'd run most of the way here. He swore it was the ghost of old Caleb Bolitho, him who murdered his wife some fifty or more years ago, then hanged himself from a tree in the wood.'

Nimbly kicking away a cockerel that was intent upon attacking a buckle on one of her shoes, she continued, 'Some of the others in here were going up to Treveddoe to have a look, but by the time they'd drunk enough to give 'em the courage they were hardly capable of walking home, let alone chasing after a ghost.' Disinterestedly, she added, 'Anyway, it was probably no more than some old dog. One like that snappy little animal the rector was in the habit of taking about with him.'

Alice was appalled by the widow's apparent indifference to the possibility that Rector Markham might have been lying all night injured in the Treveddoe woods, and something of her thoughts must have showed in her expression. Proving the years had not dimmed her perception, Widow Hodge said, 'You think I should be more concerned about the parson, dearie? If there's one thing I've learned from running a kiddleywink for so many years it's that it don't do to get involved in the affairs of anyone you're not especially close to – and I've outlived all those I cared for by a great many years. I told the parson so often enough, but it never seemed to make no difference to him. He would still keep coming around here, poking his nose into my business and pushing the teachings of his church down my throat.'

Had she not been so anxious about the well-being of Parson Markham, Alice would have taken issue with the widow about her ingratitude for the concern he had always shown towards her. However, finding the rector was far more important than arguing with someone who was too old to change her ways.

The place where the miner had heard howling could hardly be described as a wood any longer. The trees that had once existed here had been cut down for fuel when miners first came to the area. All that remained were a number of stunted, wind-deformed bushes covering both slopes of the valley, through which ran a fast-moving stream.

Unfortunately, although this might have been a quiet spot the previous evening when the frightened miner was making his way to Widow Hodge's kiddleywink, this morning the silence was shattered by the noise emanating from the Wheal Endeavour, a tin mine at the far end of the valley. Any other sound would have been drowned out by the clatter of the mine's tin stamp, accompanied as it was by the dull rhythm of a pumping engine and the intermittent screeching of pulley cables.

Alice found the din created by the mine agonisingly frustrating. She was convinced the sound heard by the miner had been the howling of Digger. Expecting the valley to be more peaceful, she had intended calling Digger, hoping to receive a response from him. Now she would have to search for the dog and his master – and she would need to take great care. Mining had been carried on in this valley for centuries. A great many shafts and deep depressions were hidden beneath lush ferns and other undergrowth.

Her search would also be hampered by the fact that the wooded hillside was traversed not by a single path but by a great many, all criss-crossing each other. None appeared to be in more regular use than another.

In spite of the noise from the mine, Alice did walk along calling Digger's name. It was as much to reassure herself that she was doing something as in expectation of receiving a response.

As she approached a fork in the path along which she was walking, she saw a man coming towards her down one of the two branches into which the path divided. He was not known to her and, assuming he was a miner, she chose the other way.

She had taken no more than thirty or forty paces when she saw something white at the edge of the path some distance ahead. A moment later she realised it was Digger.

Excitedly, Alice called the dog's name and broke into a run. Inexplicably, the dog began running away from her, at the same time barking wildly.

Calling him more urgently, Alice increased her speed. As she did so, Digger came to a sudden halt. When Alice drew nearer she was horrified to see that the small dog was standing beside a black-clad figure lying among the tall ferns growing beside the path.

She had found Parson Arnold Markham.

He lay where he must have fallen. Deeply distressed, Alice dropped to her knees beside him and took his hand. It was cold and stiff. She knew even before she felt for a pulse that her employer was dead.

Remaining on her knees, she felt tears spring to her eyes and overflow down her cheeks as Digger began whining and jumping up in a bid to lick her face. It was almost as though the small dog expected her to perform a miracle for his late master.

'Is he dead?'

The unexpected voice startled Alice. Looking up through her tears, she saw the man who had been walking towards her along the other path. Probably only a few years older than herself, he was dressed in working clothes, but, despite her distress, she realised he was not a miner.

When she nodded, not trusting herself to reply, he said gently, 'I saw you suddenly start running along the path and came after you to see if something was wrong.' Inclining his head to where the dead cleric was sprawled, he added, 'I can see something is *very* wrong. Are you related to him?'

Gathering Digger in her arms, and trying hard to regain control of her voice, Alice rose to her feet. 'I work for him, in Treleggan. When I started work this morning I realised he hadn't been home since yesterday, so I came looking for him. I was told a miner had heard howling up this way last night. I guessed it must have been Digger.'

Reaching out with the intention of fondling Digger as she spoke, the stranger drew his hand back hurriedly when the dog snapped at it. 'Well, at least the parson had a caring companion with him when he died – even if it was only a bad-tempered dog.' Aware that Alice was deeply upset, he said, more kindly, 'My name's Gideon Davey. I'm a ganger, helping to lay a new length of railway a couple of miles from here. I've just been up to the Wheal Endeavour to discuss the possibility of hiring men to set explosives for us. I'll go back there now and find help to carry the parson home. You get on back to the village and tell whoever needs to know what's happened. You'd better take the dog with you. I doubt if he'd take kindly to what we'll be doing.'

2

Gideon was an unlikely navvy. The term was an abbreviation for 'navigator', the name given to the men who traversed the countryside building railways. Despite the fact that Gideon had been so employed since he was fourteen, eleven years before, he did not have the same background as the vast bulk of his fellows.

Gideon's mother had run a small school in the west of Cornwall. His father, a naval officer, was rarely home, but his mother ensured that Gideon received a good education at her hands. Unfortunately, his father had died at sea when Gideon was still young and his mother moved to Somerset. Soon afterwards, she was married again, this time to a country parson.

Then, when Gideon was only fourteen, his mother also died. Within months his stepfather had married another widow, this time one with a family of three young girls. Gideon did not get along with his new stepmother and no longer felt welcome in what had been his family home. With everyone ranged against him, he ran away.

After trying his hand at a wide variety of temporary work, he began working on the railways that were beginning to spread across Britain. The rapidly expanding mode of transport was linking major cities and bringing town and country closer to one another. To Gideon it seemed an exciting thing to be doing.

At first, he worked for less money than the other navvies on his gang because he was too young and inexperienced to earn a man's wage. However, by the time he was seventeen he was on full pay, earning three times as much as a farm labourer could expect to take home.

An independent and itinerant breed, the navvies went wherever they were needed, boring tunnels through hills too steep for a railway to be laid over them, blasting cuttings through solid rock and laying track that would survive them, the children born to them, and their grandchildren too.

It was a hard way of life and their high rate of pay reflected the danger and lack of comfort that was an everyday part of a navvy's existence.

The rapid advance of a railway line meant that a navvy's 'home' needed to move with it. Cobbled together out of any material that came to hand, such a place was of a patently temporary nature. Only where there was a particularly long cutting or tunnel to be worked on would a brief community come into being. Anything from a few hundred to a thousand or more workers would take up residence in variously constructed shelters, each containing as many men – and women – as could be crammed into them.

In most of the larger shacks a crone was employed to wash, clean and cook for the temporary residents. Usually she would be a woman who had followed the navvies in their travels, living as the 'wife' of a series of men until whatever charm she had once possessed had been erased by drink, dissipation and the rigours of a comfortless and nomadic lifestyle.

Gideon lived in such shanties until his schooling began to pay off. Men turned to him to work out problems with their wages and protect them against dishonest 'Tommy-men' – the traders who followed the railways as they were being constructed, offering credit and charging high prices and swingeing rates of interest for shoddy goods.

He earned a reputation for honesty, in addition to having had an education. On more than one occasion he also demonstrated that he was well able to defend his decisions physically when those who benefited from the ignorance of the average navvy sought to dissuade him from giving advice that would affect their profits.

Soon, gangers too began to ask his help in working out the rate they should charge a contractor for completing a particular section of the railroad. When he came to realise how much money could be made by the leader of a hard-working gang who was able to offer a competitive rate for a particular piece of work, Gideon decided to form his own gang.

He was now in charge of fifty first-class navvies and never short of work. He could also afford to have a hut of his own, kept tidy by one of the tiny number of respectable wives who followed the railway builders. Occasionally, if the railway work was in or close to a town or village he might take lodgings, to remind himself how he had once lived.

He and his gang were currently working on the line that would extend from Penzance to a bridge being built across the River Tamar. When it was completed, Cornwall would be linked to the English railway network.

A week after the discovery of the rector's body, and not far from the workhouse where Alice had spent one of the unhappiest periods of her life, Gideon was speaking to a huge navvy who was leaning his bulk on a pickaxe. They were at the site of a cutting, being dug out of solid rock to allow a second railway line to be laid alongside the existing single track.

'Keep the men at it while I'm away, Sailor. I should have extra gunpowder and miners here in readiness for us to begin blasting first thing tomorrow morning.'

The powerfully built navvy to whom Gideon spoke had gained his nickname as a result of service in the Royal Navy. The option to serve his country had been offered to him as an alternative to prison when he had been caught smuggling, some years before.

'Since when have you needed to dress up like a tailor's dummy to go and buy gunpowder?' Sailor Smith wiped the back of a hand across his grimy and sweating forehead, at the same time grinning good-naturedly at the man who kept him and the other members of the gang in full employment.

'I'll probably be stopping off at a funeral on the way to the mine,' Gideon explained. 'I can't go into church looking as though I've come straight from work.'

'Would this be the funeral of that parson you helped carry back to Treleggan?' Sailor asked.

'That's right. He's being buried this morning. If I don't get a move on I'm going to be late.'

'I doubt if the parson is likely to take offence if you're not there on time,' Sailor commented drily. 'But I suppose that young girl who worked for him might.' More seriously, he added, 'Be careful not to upset anyone up there, Gideon. Moorland men are a peculiar lot. Anyone from much more than a mile away is looked upon as a foreigner – and they're particularly jealous of their young women. Many a would-be suitor from off the moor has suffered a painful loss of ardour when he's tried to court a moorland girl.'

'I'm merely going to the funeral to pay my respects,' Gideon retorted. 'I've only ever met this girl the once – and it was hardly a romantic meeting.'

Sailor grinned again. 'You could have said the same thing about the pretty little French girl I first met in a fish market in Brittany, but . . . All right, I know you're in a hurry. I'll tell you the story some other time – but remember my warning.'

3

By the time Gideon reached Treleggan a damp grey mist had closed in upon Bodmin Moor, bringing with it an unseasonable chill.

The funeral service had already commenced when he entered the crowded church and found a place at the end of a pew. The slight disturbance he caused was sufficient to make Alice turn round in the pew in front of him. She gave a start of surprise when she saw Gideon.

The service was being conducted by the rural dean, the Reverend Harold Brimble, who gave a eulogy on the many years of service Arnold Markham had dedicated to the parish of Treleggan.

If proof were needed that the late Parson Markham had indeed been an exceptional man, it was provided by the presence in the small moorland church of Tabitha Hodge. Despite the deprecating comments she had made to Alice about the parson, the old woman had tramped more than a mile across the moor to pay her respects to him.

Following the funeral service, the interment in the sloping churchyard was a brief affair, a niggling drizzle adding pace to the graveside ceremony.

The religious rites satisfactorily concluded, the mourners

made their way the short distance from the church to the rectory, where, with the aid of the housekeeper, Alice had set out a variety of foodstuffs, much of it donated by the late parson's parishioners.

When Gideon fell in beside Alice on the path, she asked, 'Why have you come to the funeral? You never knew the parson.'

'True,' Gideon agreed. 'But having helped carry him back here I felt somehow involved and wanted to pay my respects. I also found myself worrying about you. When we parted company I could see how upset you were – and I'd left you with that fierce animal that had been guarding the parson all night . . . where is the dog, by the way?'

Alice was not at all certain about his motives for expressing concern for her. Although she had not had a great deal to do with men, she had learned enough to realise that those who knew of her illegitimacy were apt to assume her moral standards were lower than those of other young women.

'Digger's shut in Parson Markham's bedroom,' she replied. 'He's missing his master, but he's with me for most of the day, though I don't know how long I'll be here to look after him. Reverend Brimble has asked me to stay at the rectory until a new parson comes to Treleggan. When one does . . .' She shrugged unhappily.

'What about your parents? Won't you be able to go and live with them?'

Alice looked at Gideon quickly and realised he had asked the question in all innocence. 'I have no father or mother,' she replied, offering no explanation of her statement. 'But I must go and help in the rectory now. Are you coming in?'

Gideon shook his head. 'I have business at the Wheal Endeavour.' After only the slightest hesitation, he added, 'I have to come to the mine occasionally. Would anyone object if I called in at the rectory . . . just to say hello to you?'

No man had ever asked Alice whether he might call on her, either formally or informally. For a few moments she felt ridiculously flustered. Observing this, Gideon said, 'I'm sorry,

15

Alice, I didn't mean to embarrass you. I shouldn't have asked at such a time as this.'

'It isn't that.' As she spoke, Alice looked at Gideon – *really* looked at him – for the first time. He was a good-looking young man, with strong features and eyes that were a deeper blue than any she had ever seen. 'It's just . . . I hardly know you!'

She did not add that she would probably be alone in the rectory if he called.

'That's quite true,' he conceded. 'But meeting in such unusual circumstances seemed to do away with the need for formal introductions. You know my name, Gideon Davey. I was born in Cornwall and I have no parents, either. I've had to make my own way in the world and now I'm a ganger, working on the railways with about fifty men in my gang. I negotiate the price for clearing the route for a particular section of line, then me and my men carry out the work.'

It was the barest outline of his life. Gideon did not think it necessary at this stage to explain any more.

Having had only the briefest acquaintance with moorland farmworkers and an occasional miner, Alice thought Gideon a very interesting man. She realised too that he was educated, despite his occupation.

She was also impressed to learn he was in charge of fifty men. She would never know he had needed to literally fight his way to the position he now held among the rough and incredibly hard-working navvies.

'I'm sure no one will mind you calling in at the rectory next time you pass through. You were very kind and helpful when I found the parson up at Treveddoe.'

Gideon's obvious delight gave Alice an unaccustomed thrill of pleasure.

'Thank you, Alice. I'll probably be this way again in a couple of days' time. I'll call in then.'

Alice was surprised when Widow Tabitha Hodge did not immediately return to her remote moorland cottage. Instead, after the funeral service she joined the other mourners in the rectory.

Many of the Treleggan villagers had never met the old woman, even though she lived little more than a mile from the village. Those who had were aware of her disreputable reputation. They did not want to be seen talking to her. Because of this, nursing a cup of tea and a plateful of food, Tabitha sat alone in the spacious rectory until Alice sat down beside her. Opening the conversation, Alice said, 'I didn't expect to see you here today, Widow Hodge.'

'Neither did anyone else by the look of it,' said the old woman, staring disdainfully around the room at the other mourners, who declined to meet her gaze.

'Well, *I'm* glad you came,' Alice declared. 'It would have made Parson Markham very happy to know you walked all this way to pay your respects to him.'

'I'm not so certain of that,' the widow retorted. 'But he called to see me often enough, even though I never asked him to. It was the least I could do.' Changing the subject, she said, 'The inquest found he'd had a heart attack, I believe?'

'That's right,' Alice agreed, suddenly tearful at the memory of the manner of his death. 'Thankfully, he would have died instantly. I would hate to think he'd been left lying out there all night, suffering. As it is, it's thanks to you that I found him so quickly.'

'I've heard it said you weren't on your own when you found the parson's body. That you had some navvy with you.'

'I didn't have him *with* me,' Alice said firmly. 'He was coming along the path from the Wheal Endeavour when I saw Digger and began running. He came to find out what was happening, then went back to the mine to get help to bring Parson Markham here.'

Widow Hodge nodded her head slowly. 'I thought it would be something like that but you know what village folk are like. If there's nothing for them to gossip about, they'll make it up.'

She made this observation in such a loud voice that more than one woman in the room felt the comment was directed at her.

'Talking of people causing trouble . . . was that the same

17

navvy you were talking to on the way from the church? The one who was at the service?'

Alice nodded. 'He had to visit the Treveddoe mine again and decided to come to the service – but he's a ganger, not just a navvy.'

'He's trouble, whatever name you care to give him,' Tabitha declared positively. 'Not that it'll be of his own making if it comes, I dare say.'

'What do you mean?' Alice demanded indignantly. 'All he's done is helped bring poor Parson Markham back to the rectory and come to a service to pay his respects. That's not making trouble!'

'As long as that's all he ever has to do with Treleggan no doubt folk will soon forget all about him,' said Widow Hodge. 'But if he shows his face here again, those gossips I was talking about will make a scandal of it – and it won't take much to cause mischief. Billy Stanbeare and one or two of the other young Treleggan men saw you and this navvy talking and they were taking more than a passing interest. It wouldn't be the first time a girl choosing a man from outside her own village has caused trouble. Men like Billy tend to take it as a personal slight.'

Billy Stanbeare was the son of Henry Stanbeare, the church-warden who had been most bitterly opposed to Alice's taking work in the rectory. Billy was also a moorland wrestling champion – and a bully. He and his friends enjoyed reminding Alice of the circumstances of her birth whenever an opportunity arose.

With this in mind, Alice said bitterly, 'I would have thought Billy and his father would be more pleased than anyone else if someone were to take me away from Treleggan. Billy has told me often enough that the village doesn't want the likes of me living here – and his pa never forgave Parson Markham for giving me work at the rectory.'

'Ah! There you've put your finger on the root of your troubles,' Tabitha Hodge said sagely. 'It's Henry Stanbeare – and always has been. I doubt if you've heard the story, but when Billy's mother died Henry wasted no time looking around for someone to take her place. It should have been no trouble for him. After all, he has his own farm, is a churchwarden and likes to think of himself as a respected member of the community – a cut above everyone else. The woman he chose was your mother. Trouble was, he began to tell others of his plans before he'd spoken to her about 'em. She was the last to know what he had in mind – and he was the last to know she already had you in her belly. As a matter of fact, when you were born there were many who seriously thought *he* was your father! A few of 'em even refused to speak to him when your ma went to prison for not naming him. Of course, it all came out when your real father died, but Henry was made to look a fool. It stuck in his craw. He never forgave your ma for that. From what I hear he's never forgiven you either.'

Aware of Alice's astonishment, and satisfied she had not known the story, Tabitha took a noisy sip of her tea before saying, 'It's men like Henry Stanbeare who made me decide years ago that if he represents what having religion is all about, I'd rather do without it.'

Once more her voice had risen as she spoke and heads were turned towards her, including that of the subject of her criticism.

Recovering from Tabitha's revelation, Alice pointed out, 'Yet you've come to Parson Markham's funeral.'

'I came out of respect for the man,' Tabitha declared, 'not what he stood for. Now I've seen him put in the ground it's time I went home. There's work to be done and it won't do itself.'

Tabitha's age showed when she struggled to rise from the wooden-armed chair she occupied but, brushing aside Alice's helping arm, she eventually succeeded by herself.

Alice accompanied the indomitable old woman from the house and walked with her as far as the gate that led from the rectory garden. Before bidding the widow farewell, she said, 'I'm glad you came to the funeral, Widow Hodge – and thank you for telling me . . . telling me about Henry Stanbeare and my mother. I've never really understood why he's always hated me so much. I'd like to call on you now and again. Until we get a new rector I'm not likely to be too busy. I could perhaps do some cleaning, washing or ironing for you?'

'I'm quite capable of looking after myself, thank you, young lady. Because I came to pay my last respects to the parson doesn't mean I'm going to let anyone else poke into my affairs. Besides, you've had to work hard all these years to prove you're not the sort of girl the gossips say you are. Don't throw your reputation away now by coming to visit me – and remember what I've said about that young man of yours. If he comes calling on you he'll be walking into trouble.'

With this, Tabitha Hodge walked away, heading for a path that passed through the churchyard and on to the moor beyond, leaving Alice staring after her. She was astounded that the old woman should know so much about her – and be familiar with her reputation among the Treleggan villagers.

4

I

Tabitha's warning was passed on to Gideon later that same day – not by Alice, but by the aged widow herself.

Gideon had satisfactorily negotiated the purchase of a quantity of gunpowder from the captain in charge of the Wheal Endeavour and enlisted the services of a 'pare' – or team – of six men to help his gang blast away the solid rock they had encountered in the cutting on which they were working. A number of the navvies, Gideon included, were familiar with the use of explosives, but they were not experts. These miners were.

The 'pare' consisted of some of the most experienced men employed in the mine. They would work for Gideon in teams of two, twenty-four hours a day, in addition to carrying out their work in the mine. Making use of their expertise would minimise the risk of accidents. It would also save time, and for navvies time was money.

With agreement reached on satisfactory terms for all involved, the spokesman for the pare suggested they should celebrate with a drink or two at Widow Hodge's kiddleywink before going their separate ways.

The men arrived at the remote cottage somewhat earlier than

the widow would normally expect to receive customers, and she was alone in the kiddleywink. However, too shrewd a businesswoman to turn them away, she invited them inside and served them herself.

The heyday of the Cornish smuggler had passed many years before, yet it was doubtful whether excise duty had been paid on the drinks served out to those who frequented Widow Hodge's establishment.

By the time others had drifted in from the moorland mines, two serving-maids had taken over from Tabitha Hodge. Named Florence and Fanny, the two girls were sisters who needed to work in order to support their mother and seven younger brothers and sisters. The family had been left destitute when their miner father deserted them to try his luck in the mines of Mexico. That had been two years before. Nothing had been heard of him since.

Desperate for money to support their large family, the girls had an 'arrangement' with Widow Hodge that was satisfactory to all the parties involved. The girls worked without pay. In return, she accepted that one or other of the sisters would be absent for much of the night in order to 'entertain' a series of miners in the hay barn behind the kiddleywink.

Tabitha Hodge had recognised Gideon as soon as he came in, but she said nothing to him until Florence and Fanny had taken over the serving chores. Then, carrying a glass of brandy drawn from a smaller keg than the ones which served her less discerning customers, she crossed the room and seated herself at the table where Gideon and the men from the Wheal Endeavour were drinking.

Taken aback by her uninvited presence among them, Gideon was even more surprised when she addressed him.

'Treleggan folks must have wondered what a navvy was doing at the funeral service for their parson today, and some of the young village men were none too pleased to see you paying attention to Alice Rowe. Billy Stanbeare in particular.'

Recovering from his surprise, Gideon replied, 'I didn't see any of them out on the moor searching for the parson when he'd been missing all night – but I did see Alice. Besides, it's none of their business. I've never felt obliged to ask permission to speak to anyone I please.'

'I'm not suggesting you should.'

Tabitha raised the brandy to her lips. Taking a sip, she studied Gideon over the rim of the goblet. When she lowered it to the table, she added, 'All the same, you'd do well to take heed of what I'm saying. The ways of the rest of the world haven't yet reached Treleggan. I sometimes wonder whether they ever will. That Billy Stanbeare, in particular, is one to watch. He's got a nasty streak in him. Inherited it from his father.'

Gideon was about to protest that he had no intention of being dictated to by a village bully, but Tabitha held up her hand to silence him. 'I know, you're a navvy. You don't give a damn for anyone and in another few weeks you'll be moving off some-where else. Once you've gone Treleggan might never cross your mind again – but Alice Rowe will still be in the village. She's had plenty to put up with from the Stanbeare family over the years. The last thing she needs is for you to open up old wounds and stir the Stanbeares up against her again.'

'Is this the girl whose mother went to prison for refusing to name Doug Mitchell as the father of her child, after he'd been hurt at the Wheal Endeavour?'

The question came from one of the mine's shift captains who had come in to join Gideon and the miners.

'That's right,' Tabitha replied. 'You'd have been up at the mine at the time when he was hurt.'

'I worked with him,' said the shift captain. 'He was a good man. He and the girl's mother – Grace, I think she was called – would have married had it not been for the accident. Mind you, they had need to keep their courting quiet – for the very reason you were just warning Gideon about. Them at Treleggan are farmers, or farmworkers. We're not. We're outsiders and they won't tolerate us going near their women.'

'Tell me more about Alice's mother and father.' Gideon saw this as a good opportunity to find out more about Alice. He was not disappointed. The shift captain told all he knew, with

23

Tabitha occasionally interrupting to fill a gap in his knowledge.

By the time they finished talking, Gideon had built up a vivid picture of the unhappy life Alice must have led in a village where she was ostracised by everyone. It might not have been a complete record of her life but the shrewd old widow was surprisingly well informed about what went on in Treleggan.

Eventually, Tabitha gave Gideon an enigmatic look before asking, 'Have you learned enough to decide not to see Alice again?'

Gideon shook his head. 'I don't intend allowing this Billy Stanbeare or anyone else to tell me what I can or can't do. But I promise you that he and his friends will be left in no doubt about what's likely to happen to them if they vent their spite on Alice when I have to move on.'

Tabitha gave a resigned shrug of her shoulders. 'I hope they take notice of you, but I doubt it very much. Right now they're full of themselves and the good harvest the farmers around here have enjoyed. That reminds me . . .'

Turning her attention to the shift captain, she said, 'Warn your men not to come here on Saturday of next week. Billy Stanbeare and the farmworkers are coming to celebrate harvest-home. They'd enjoy cracking a few miners' heads as part of the celebrations – and I've no doubt my kiddleywink would suffer in the process.'

'I'll tell the men,' the shift captain promised. 'It'll make a change for some of their wives to have them home on a Saturday night.'

II

The warning to Gideon about the danger posed by the young men of Treleggan came too late for him to avoid trouble.

His route back to the railway line took him through Treleggan village and he was hurrying because he wanted to reach the camp before dark. He had thought of calling in on Alice, but

24

decided against it. He felt it was probably too soon after the funeral.

He had passed out of sight of the small cluster of village houses when he saw a group of perhaps twenty young men coming along the lane towards him. Among them he recognised one or two who had been in the church during the funeral service for Parson Markham that morning.

From the noise the men were making he realised they had been drinking. In fact, although Gideon was not aware of local farming practices, neighbours were in the habit of helping each other out at harvest times and these farmworkers had just completed gathering in a bumper crop from a local holding. Their reward had included a generous supply of very alcoholic home-made cider.

When they spotted Gideon there was much nudging and whispering among them. He would rather not have met with them, but there was nothing he could do about it now. He hoped he might be able to pass through the group with no more than cursory greetings being exchanged.

As he drew nearer he realised this was not going to be possible. The men had deliberately formed a line to effectively block the narrow lane.

His hopes fast fading that he would be allowed to pass on his way unmolested, Gideon knew he must try.

Reaching the Treleggan men, he headed for a small gap between two of them. Before he reached it a squat, broad-shouldered man with long, black, untidy hair and a matching beard stepped into his path.

'I suppose you've been to Treleggan again, to see her up at the rectory?' The bearded man's voice was in keeping with his appearance.

'No,' Gideon replied, 'I've been to the Wheal Endeavour.' As he spoke, he was aware that the bearded man's companions had closed in around him, making any attempt to run from them quite impossible, even had he been so inclined.

'He speaks well for a navvy and a foreigner, don't he, Billy?' said one of them, addressing the bearded man.

'Most navvies are able to speak,' Gideon retorted. 'And many are not foreigners. They're Cornishmen, same as me.'

'Ah! But you're not from round here,' Billy Stanbeare pointed out. 'That makes you a foreigner as far as we're concerned – and we don't take kindly to your sort coming in after Treleggan girls.'

'I'd hardly call carrying the body of a dead parson back to the rectory "coming in after a girl",' Gideon replied. 'I'd say it was being neighbourly.'

He spoke quietly and evenly. Aware that the Treleggan men had been drinking, he did not want to antagonise them unnecessarily. Yet he realised it would be unwise to show any sign of fear.

'It seemed to me you was a bit *too* neighbourly when I saw you talking to Alice Rowe after the funeral this morning,' Billy Stanbeare said belligerently.

'The rural dean spoke to her for far longer than I did,' Gideon retorted. 'Will you be having words with him too?'

He held his breath for a moment, realising he had reacted too sharply to the other man's words.

Billy Stanbeare looked at him uncertainly for a moment, and Gideon might have got away with his indiscretion had not one of the farmworkers sniggered. Billy's eyebrows came together angrily and Gideon knew he was in trouble.

'We've got a clever one here,' Billy said. 'I think he should have a taste of what'll happen to him if we see him sniffing round a Treleggan woman again – grab him.'

Gideon knew that if he was going to put up a fight, now was the time to do so. He also suspected that this was what Billy was hoping he would do. With odds of twenty to one against him, Gideon was aware he would be battered insensible within minutes – if he were fortunate enough to suffer nothing worse. He made no resistance when the men grabbed him and pinned his arms to his side.

As he had suspected, Billy appeared disappointed he had not put up a fight. The farmworkers' leader said, 'We'll give him a ride off the moor, shall we? Same as we did that Liskeard teacher who came courting Nellie Collins.'

His companions cheered the proposal and one said, 'We'll find a good mount for him down at the spinney alongside Jan Thomas's lower field. Let's take him down there.'

The men were in high spirits as Gideon was marched along the lane and led into a field. They took him to a spinney where a number of trees had been chopped down – and here they chose his 'mount'.

It was a tree trunk, roughly trimmed of branches and not much thicker around than his upper arm. Gideon was forced to straddle it, and then it was raised from the ground by four men, who tucked it beneath their arms.

When the men set off with him on the log, the movement sent a sharp pain right through his body, but it was immediately apparent that not enough of his tormentors could stay sufficiently close to keep his arms pinned to his sides. Breaking free of them, he grasped the tree trunk in order to support his weight on his arms.

Some of the men tried to wrest his hands free, but Billy said, 'Leave it. He won't be able to hold himself up like that for very long. A couple more of you help with the weight of the pole. It only needs two to walk alongside him to stop him from falling off. The rest of you stay close and make certain he doesn't get away – not that he seems inclined to give us any trouble, and by the time we're finished with him he'll be lucky to walk, let alone run. He certainly won't be much use to Alice Rowe for a while.'

Billy had not taken into account the arm muscles Gideon had built up during years of navvying. Although he was now a ganger, he still worked alongside his men on most days. Even so, he was unable to hold his body clear of the pole for the whole of the time he was being jogged along, especially when Billy ordered the men supporting the trunk to break into a trot.

However, when the farmworkers arrived at a spot where the path wound its way up a steep hillside, they were forced to slow down, the pole-bearers complaining to their leader that some of the others should take their turn carrying the tree trunk with its burden.

At the same time, due to the slope and the narrowness of the path, the ring of men surrounding Gideon were forced to either go ahead or stay behind the pole and its reluctant rider, leaving only two men alongside him.

Now, thanks to the uneven slope, Gideon's feet would occasionally touch the ground – and he realised this gave him the chance of escape for which he had been waiting.

At a moment when the pole-bearers were complaining particularly vociferously as they struggled with their burden, Gideon's feet touched the ground once more. Releasing his grip on the trunk, he clenched his fists and flung his arms up and backwards, knocking the two supporting farmworkers off their feet.

Vaulting off the pole, Gideon turned a number of somersaults as he tumbled through gorse and fern down the steep hillside. When he picked himself up at the foot of the hill, he found he was on a narrow track, made over very many years by the hooves of moorland sheep. The track curved away to the right, while the higher path on which the angry farmworkers were standing went to the left.

Pursued by the cries of the men from whom he had escaped, Gideon began running. He did not ease his pace until the sound of shouting, mixed with contemptuous laughter now, died away in the distance.

5

I

On the day after the funeral of Parson Arnold Markham, word of what had befallen 'Alice's navvy' was known to everyone in Treleggan within an hour of work's beginning.

To everyone, that is, except Alice, who was the subject of more than the usual number of sly looks from the women she passed as she walked to the farm at the edge of the village from which she purchased milk.

As the farmer's wife measured out the milk, she said casually, 'After what went on last night I don't suppose we'll be seeing that young man of yours in the village again?'

Puzzled, Alice said, 'Young man of mine? What young man?'

'Why, him who came to the funeral yesterday. The one you were with when you found the parson lying dead up at Treveddoe.'

'No one was *with* me when I found Parson Markham,' Alice retorted. 'If you're talking about the navvy from the railway who helped carry poor Parson Markham back to the rectory, he came along *after* I'd found the parson. I'd not met him before then. But what's happened to him?'

'Seeing as you don't know him you've no need to concern

yourself about him.' The disbelief of the farmer's wife was evident. 'But my Albert said this navvy got himself in trouble because he'd come to Treleggan to see you – and Albert had no reason to lie to me. He said he and some of the others were coming from harvesting at the Coumbe farm when they met up with this young man of yours. He was on his way back to the railway from the village. Him not being one of us they warned him about coming to see a Treleggan girl. When he didn't seem inclined to take notice of 'em they gave him a ride on a wood-cutter's donkey – one of the trees that had been cut down in Jan Thomas's spinney.' The farmer's wife chuckled. 'I've never heard of anyone coming back for another such ride, so I reckon you've seen the last of him. A navvy, you say he is? Then it's a good thing for all of us that he's gone. We can do without the likes of them around Treleggan.' Giving Alice a sly look, she added, 'But, of course, you not knowing him, it won't matter to you one way or the other, will it?'

On the way back to the rectory, Alice fought hard to contain the anger and frustration she felt. She was also concerned for Gideon. She tried to convince herself that such feelings were quite ridiculous. She had only met Gideon twice and neither occasion could be considered even remotely romantic. However, Gideon *had* asked if he might call on her and she had fallen asleep the night before trying to analyse the reason behind his request.

She realised she had probably read far too much into what Gideon had said to her. Even so, it had been the closest she had come to any sort of romance in the village which had ostracised her for so long.

Gideon's humiliation was a source of amusement to the young men of Treleggan for only a few days before it was replaced by anticipation of the harvest-home celebrations that would take place on the following Saturday.

They boasted among themselves of the prodigious amounts of alcohol they would consume during the evening. When mention was made of Florence and Fanny, the two sisters who worked at Widow Hodge's kiddleywink, the young men of

Treleggan vowed the two serving-girls would enjoy a busier night than ever before. They declared that far more of their time would be spent in the hay barn than in serving drinks.

When the day arrived the celebrants met at the kiddleywink early in the afternoon. It promised to be a memorable occasion. The farmworkers had received their pay, while the farmers' sons had extracted an unusually generous allowance from their fathers. It was a rowdy mob of about thirty young men who crowded into Widow Hodge's moorland cottage and they became noisier as the day progressed.

Then, later that evening, an ashen-faced Fanny hurried inside the kiddleywink to be greeted by cries of 'Who's next?'

However, as the serving-girl's agitation became apparent, the farmworkers fell silent and Tabitha demanded, 'What's happened, Fanny? Where's the lad you went outside with?'

Instead of answering, Fanny searched the room with her eyes until they fell upon Billy Stanbeare, and it was to him she spoke.

'I was sent to find you – by a navvy. Him as was in here a few days ago.' Shifting a frightened gaze to Tabitha, she explained, 'It's the navvy who was here with some of the miners from Wheal Endeavour on the day of the parson's funeral.' Turning back to Billy Stanbeare, she added, 'He said I was to find you and tell you he wants to speak to you outside.'

Letting out a loud whoop of glee, Billy called to his companions, 'It's Alice Rowe's friend! He'll no doubt want us to change our minds and let him visit her. Come on, we'll show him we meant what we said – and we'll make certain he doesn't run away this time!'

Billy led the rush of noisy men to the door, the others jostling in an effort not to miss the 'fun' they were convinced was about to begin.

Those who were first out of the door came to a sudden halt, drawing protests from those still trying to leave the kiddley-wink. Only when the young farmworkers finally struggled outside did they see the reason why there had been a sudden halt and they fell silent.

Gideon was outside the kiddleywink – but so too were the

31

navvies of his gang, some fifty men, most armed with hickory-wood pickaxe handles.

'Come away from the door. We don't want to have to hunt you down in Widow Hodge's house and cause any damage, do we?'

Gideon's warning proved unnecessary. Accustomed to the fights that occasionally broke out among her customers, Tabitha had followed the farmworkers to the door. One glance at what awaited them was enough. She slammed the door shut, rammed the bolts home and ordered the two serving-girls to clear the tables and wash and dry the glasses. Whatever the outcome of the confrontation outside, when it came to an end she would throw open the door and suggest the victors came in to celebrate – but she doubted whether they would be serving drinks to local men.

Aware they had no means of escape, the farmworkers crowded together, looking apprehensively about them at the semicircle of tough navvies.

'What's the matter?' Gideon taunted them scornfully. 'Don't you like the odds today? It's not even two to one! But I seem to remember you prefer it to be twenty to one – in *your* favour. Never mind, I'm not a vindictive man . . . well, perhaps just a little.' Addressing Billy Stanbeare, he said, 'I believe you consider yourself as the leader of this little lot? At least, that's how it seemed to me when I was waylaid by you and your friends. I also hear you fancy yourself as a wrestler? Well, I'm no wrestler, and I don't fight by any rules, but I'm willing to take you on – just you and me – without involving the others.'

There was a gasp of disbelief from the Treleggan men. Billy had been the undisputed wrestling champion of the area for a couple of years, but because he enjoyed hurting those with whom he wrestled he now found it difficult to find opponents. The listeners could hardly believe Gideon would choose to go against him in a 'no rules' grudge fight.

'Well, are you going to take up my challenge, or do you need twenty or thirty men behind you before you have the courage to fight a navvy?'

Instead of replying, Billy pulled off his coat and threw it to one of the Treleggan men.

Gideon was doing the same when Billy charged at him. Although he still had one arm in a coat sleeve, Gideon had been half expecting such a move. The farmer's son ran on to a heavy boot, planted painfully in his lower stomach.

'You'll need to be quicker than that, Billy,' Gideon taunted and was ready when Billy made another rush at him, arms held wide in an attempt to grasp him in a wrestling hold. This time Gideon ducked to one side and swung a well-aimed punch that caught the other man full in the face.

Billy dropped to his knees with a grunt of pain. When he rose to his feet, shaking his head, his nose was clearly broken and bleeding profusely. It set the pattern for a grim, unremitting battle that lasted for a full fifteen minutes to the noisy approbation of the navvies.

Billy's supporters watched in disbelieving silence as their champion repeatedly rushed forward with increasing desperation, only to be met by a heavy fist or, occasionally, a kick. Only once did he manage to secure a hold on his agile opponent, but he staggered back immediately, screaming he had been blinded by Gideon's two-fingered jab.

It developed into such a one-sided contest that the navvies were almost as relieved as the farmworkers when Gideon brought it to an end. Stepping close to the staggering and virtually helpless Billy, he landed two punches that put him on his back on the soft moorland turf, arms extended at his sides, totally unconscious.

Gideon had not quite finished with him. Pointing to a couple of men at the front of the group of farmworkers, he ordered, 'You . . . and you. Take him to the duck-pond and throw him in. I want him conscious to hear what I have to say . . . *Do it now!*'

He barked the last three words when the men were slow to move, and, hurrying forward, they dragged Billy to the edge of the pond. By the time they reached it Billy was coming round, but Gideon said, 'Throw him in.'

Used by Widow Hodge's geese and ducks, the pond was fed by a sluggish stream and was extremely muddy. It was also quite deep and for a moment the two men hesitated. However, when a number of the miners advanced towards them eagerly, the unhappy farmworkers took hold of their semi-conscious friend and pitched him clumsily into the water.

Billy disappeared beneath the surface and for a moment navvies and farmworkers held their breath. Then he rose to the surface coughing and spluttering, floundering in the mud in a clumsy attempt to stand up.

Gideon turned to the subdued Treleggan men. 'Now go in there with him and see that he comes to no harm.' When they looked at him in disbelief, Gideon addressed his gang. 'Drive them into the pond. Use your pick handles if you need to. A few cracked heads might help them remember what I'm going to say to them.'

Pick handles were formidable weapons in the hands of the navvies. Outnumbered, the Treleggan men stampeded towards the pond and splashed their way through the cold water towards their still-choking companion. Soon they were all huddled together about Billy, waist-deep, the mud of the pond gripping their legs as high as their knees.

'Have you paid for the drinks you've had in the kiddley-wink?' Gideon asked them.

Heads were nodded in confirmation and one man added, 'There's money in there for drinks we haven't had, too.'

'That should pay me for a suit to replace the one you ruined the last time we met,' Gideon retorted. 'Now, is Billy Stanbeare sensible enough to hear me?'

Heads turned to look at their leader and one of the men called out, 'He can hear you.'

'Good! I want him and every one of you to take in what I have to say.'

Gideon moved closer to the pond and raised his voice so that his words would carry to each of the uncomfortable farm-workers.

'The last time I came to the Wheal Endeavour you rode me

34

off the moor on a tree trunk. What's happened to you today is as nothing compared to what will happen if you try to do anything similar again. I'll come and go as I wish, whether it's to here, the Wheal Endeavour – or Treleggan. If any one of you gives me any trouble – any trouble at all – you'll get what Stanbeare has been given, only worse. What's more, you won't take out your spite on anyone else – and I don't think I need mention any names. If I find out you've done anything to make life difficult for *anyone* because of me then I and my gang will come back and it won't only be your bodies that will suffer. My men will bring their picks with them. They can demolish a house in less than twenty minutes – a village the size of Treleggan in a day. That's my promise to you – and don't think you'll be able to wait until I've moved on to do as you like. There will be navvies working on the lines in this part of the world for the next ten years, or more. Word will be passed on and they'll be just as eager as my men to enjoy a little fun at the expense of a house or two. Have I made myself perfectly clear?'

The farmworkers responded with a disgruntled murmuring among themselves and Gideon repeated his question.

'I said, have I made myself clear?'

This time their affirmation was more audible.

'How about you, Billy? In the event of trouble from *anyone* your house will be the first to be knocked down, so I want to be quite certain you understand.'

'I heard you.' Billy mumbled the words painfully through broken and swollen lips.

'Good!' Satisfied his victory was complete, Gideon said, 'Now I suggest you all go home and change into dry clothes. There's a cold wind getting up. You don't want to end up with a chill.'

The Treleggan men extricated themselves from the pond with some difficulty and, filing past the grinning navvies, set off for their homes. When they were well on their way, Gideon looked round at his men and said, 'Well, I don't know about you, but I'm about ready for a pint of ale.'

Closely following events from a window, Widow Hodge

hurried to open the door wide as Gideon said, 'Come on. The first round's on Billy Stanbeare and his friends, the second is on me . . .'

II

'I heard what you said to Billy Stanbeare and his friends. What makes you think they'll take any notice when they get over the scare you've given 'em today?'

Tabitha Hodge put the question to Gideon when he was seated at a table in the kiddleywink, drinking with Sailor and some of the other navvies.

'They'd better. I once saw a gang of navvies take twenty minutes to knock down the house of a man who'd done them out of money he owed them. My gang are more skilled and twice as loyal as they were.' Setting his glass down on the table, Gideon added, 'Do *you* think I convinced them?'

'I'd say most believed you. I'm not certain about Billy Stanbeare. You should have thrown him in the pond while he was still unconscious and put an end to his nonsense once and for all. Mind you, he'll not cause any trouble until he gets some of his pride back. It's a great pity you didn't happen along years ago. You'd have been able to save Alice – yes, and her mother too – from suffering a lifetime of grief.'

'You're fond of Alice, aren't you?' Gideon was puzzled by the depth of feeling Tabitha was displaying on Alice's behalf. He had only a passing acquaintance with the owner of the kiddleywink, but this, coupled with what he had heard of Widow Hodge from the miners of Wheal Endeavour, had left him with the impression of a hard woman. Her concern for Alice seemed out of character.

'She's very like her mother was at her age,' Tabitha said, avoiding giving him a straight answer. 'She cares too much for others. In the main, folk just aren't worth caring about.'

Gideon felt there was much about Widow Hodge he did not

know and he doubted whether she would ever tell him, or anyone else.

'I've warned Stanbeare what will happen if he makes life difficult for Alice – but I need to know if he does. Will you pass the word on to me or one of the other gangers if he gives her any trouble?'

'I'm a mile away from Treleggan, and a great deal farther away from the railway. Even if I heard of something going on I wouldn't know how to get word to you. Besides, why should I take sides? The reason my kiddleywink is open when others have come and gone is because I mind my own business. I'm too old to change my ways now.'

'That's reasonable enough,' Gideon conceded grudgingly. 'It's just that I would hate to feel I've made things worse for Alice, instead of better.'

'I doubt if things will get any better,' said Tabitha, 'but there's a whole lot of difference between disapproving of someone and doing 'em harm. Them who live in Treleggan have a lot to answer for in the way they've treated that girl, but I doubt if they'd condone out-and-out violence against her, not even from Billy Stanbeare.'

'I hope you're right,' Gideon said. 'But I'd still like to feel there was someone who would let me know if things did get worse for her.'

The old woman sniffed noisily before saying enigmatically, 'I expect you'll hear about it if Alice is picked upon.'

'I trust I will.' Gideon downed the remainder of his drink before saying, 'I'm going now. I'll leave you with money for another round of drinks for my men. After that they'll be buying their own.'

'I'll make certain they know that,' Tabitha declared. 'But as you've run off all my other customers I'll take enough money from you to pay for at least *two* rounds. After that they can stay for as long as they like.'

Taking money from his pocket, Gideon grinned. 'I reckon I owe you that – and you won't lose anything by having my gang as customers instead of the Treleggan men. They've been paid today and will keep you and your two girls happy for as long as you're serving them. All I ask is that you throw them out if they are still drinking this time tomorrow. I want them working first thing on Monday morning. I too have a living to make.'

'No one with money to spend has ever been thrown out of my kiddleywink,' Tabitha retorted. Changing the subject, she asked, 'Are you going straight back to the railway?'

Puzzled by her question, Gideon said, 'I'd like to call in on Alice and tell her what's happened here tonight. But it might not be such a good idea. It's dark now. She might already have gone to bed.'

'She'll be up and about,' Tabitha said positively. 'I've got a few cooked chickens not long out of the oven. There'll be plenty for your men and you can take a couple with you. Drop one in to the rectory. It'll give you an excuse to call there – if you feel you need one.'

Surprised by her apparent generosity, Gideon asked, 'Do you want me to pay for them?'

'They cost me nothing,' Tabitha replied. 'The Treleggan men brought them to eat during the evening. I doubt if they'll be back asking for them and it would be a pity if they were wasted.'

Gideon suspected the unfathomable old widow was deliberately giving him a reason for calling upon Alice. He could not think why she should, but he would not refuse the offer.

III

Although Gideon doubted whether any of the Treleggan men would be looking out for him, he was glad of the darkness when he entered the village.

Turning in at the rectory gate he paused in the shadow of some tall elm trees for a few minutes, making certain he was not being followed. From where he stood he could see lamps burning behind curtains in at least two of the downstairs rooms, so he knew Alice had not gone to bed.

Satisfied no one from the village was watching, he approached the house. Tugging at the bell-pull, he heard a faint ringing. It was followed by the sound of a dog barking somewhere inside the large building. He assumed the dog was Digger.

It was almost a full minute before Alice's querulous voice from the other side of the door called, 'Who is it? What do you want?'

'It's Gideon.' He kept his voice low so it would not carry beyond the rectory garden. 'I'm on my way back to the railway but have been given a couple of cooked chickens and told to drop one in to you on my way.'

There was a long silence from within the rectory before he heard bolts being drawn and the door swung open.

'Who would be sending food to me?'

Alice was standing in an unlighted hall but a sliver of light escaping from a partly open door farther along a passageway was sufficient for him to see she was holding Digger in her arms.

'It's from Widow Hodge . . . but there's a story attached to it.'

There was another brief hesitation before Alice said, 'You'd better come in for a few minutes. We don't want anyone from the village seeing you here.'

Entering the hallway, Gideon said, 'I don't think anyone will be looking for me to come here tonight. The young men in the village have had all the trouble they can handle for one day.'

As he followed Alice to the kitchen he outlined what had occurred at Widow Hodge's kiddleywink and the source of the chickens he was carrying.

Alice seemed concerned. 'You fought with Billy Stanbeare? But . . . he's a champion wrestler.'

'So he may be,' Gideon replied, 'but he has a lot to learn about fighting.'

Aware that talk of being involved in a fight might give Alice the wrong impression of him, he added quickly, 'Stanbeare had me ridden out of the area on a tree trunk a few days ago. Had

39

I not taught him a lesson I would never have dared show my face in Treleggan again.'

They had reached the kitchen now, and seeing Alice's concerned expression Gideon thought she might be thinking about the implications for her.

For a few moments there was a distraction in the form of Digger. Unlike the last occasion they had met, when Gideon had tried to stroke him, the dog now behaved as though they were life-long friends.

When Alice called the dog away and made him lie down in his basket, placed on the floor in a corner of the room, Gideon said, 'I told Stanbeare and his friends that if they try to vent their spite on anyone associated with me in any way I'll come back with my gang and reduce the houses of those involved to a heap of rubble. They were all standing up to their waists in the mud and water of Widow Hodge's duck-pond at the time, so I don't think they're likely to forget my warning.'

'Short of actual violence there's not a lot more they *can* do to me,' Alice commented bitterly. 'Billy and his father have been largely to blame for the way I've been treated. It's high time one or other of them had his come-uppance.' Suddenly giving Gideon a direct look, she asked, 'Why should you *want* to show your face in Treleggan again? The railway has no business here.'

'I'm not used to having men like Stanbeare dictate what I can or can't do. It seemed to me it was time someone showed his friends that he's not as tough as he thinks he is.'

'Oh! So it was your pride that brought you back to Treleggan?' Alice was uncertain whether her feelings were those of disapproval, or disappointment.

'Pride was certainly involved,' Gideon admitted. 'But I wanted to see you again and I wasn't going to let Billy Stanbeare – or anyone else – stop me.'

Giving him a rather disconcerting direct look once more, Alice said, 'We've only met the once – twice if you include the few minutes we spent chatting after the funeral. What was so important about seeing me again?'

'I've asked myself the same question,' Gideon replied, his

honesty surprising him as much as it did Alice. 'I couldn't think of an answer, so I thought I'd better call on you again to see if I could find one.'

Gideon thought he might have said too much, but when Alice spoke again it was to say, 'Would you like a cup of tea? The kettle's boiling.'

'Yes, thank you. There's quite a cool wind blowing outside on the moor.'

'There usually is,' Alice commented, 'even at this time of the year.' She was surprised by the ease with which they could talk to each other after knowing each other for such a short period of time.

When she put a cup of tea on the kitchen table in front of him, she noticed with concern the cuts and grazes on his knuckles.

'Are your knuckles hurt as a result of fighting Billy?' she asked.

When Gideon nodded, Alice realised it must have been a far more violent encounter than she had at first thought.

'Is Billy badly hurt?'

'He has a broken nose, but he'll live.'

Seeing Alice looking at him wide-eyed at the thought of the fight, Gideon said, 'Do you care? From what Widow Hodge told me he's made your life miserable the whole time you've been here, at the rectory.'

Alice nodded. 'If it hadn't been for Parson Markham I couldn't have stayed in the village.'

Mention of the late parson brought unwonted tears springing to her eyes and she turned away, but not before Gideon had seen them.

'I'm sorry, Alice. I didn't come here to upset you.'

'You haven't.' When she faced him again Alice made a brave attempt at a smile. 'I haven't yet got used to the fact that I've lost him. I have a little weep at least two or three times a day.'

'From what I've heard from Widow Hodge, he was a very good man,' Gideon said. 'I have the impression she doesn't waste such praise on many people.'

'She would never praise him to his face,' Alice said, 'but it's clear now that she had a sneaking regard for him.'

As they were speaking, Gideon picked up the cup of tea Alice had made for him. The act of putting his index finger through the handle caused a cut on the knuckle to break open.

Alice saw it and said, 'I'll put something on those knuckles for you before you have your tea.'

'It's all right. Working on the railway you get used to cuts and grazes.'

'I'll still put something on them, otherwise you'll be dripping blood over the table and floor and I'll need to scrub them clean.'

Gideon made no further argument. He allowed Alice to clean his cuts before applying an ointment she took from a jar kept in a kitchen cupboard.

Alice tried not to look at his face as she dressed his grazed knuckles. She found the act of holding his hand in hers quite disturbing enough.

When she had done, Gideon squeezed her hand momentarily as he said, 'Thank you.'

There was a silence between them for a few moments before Gideon asked, 'Are there other servants living in the rectory?'

Alice became immediately wary, suspicious of his reason for asking such a question. 'There's a housekeeper,' she replied, 'but she's not in at the moment.'

Gideon looked concerned. 'I'm sorry. I shouldn't really be here if you're alone in the house.'

Aware she had jumped to the wrong conclusion, Alice said, 'If it's my reputation you're worried about, don't waste your time. The people of Treleggan decided the sort of person I was on the day I was born.'

'All the same . . .'

'All the same *nothing*!' Alice said.

Suddenly aware that she was behaving in a far more posi- tive manner than she could ever remember before, Alice became concerned it might frighten Gideon into leaving. Then, looking once more at his bloody knuckles, she decided he was not a man who frightened easily.

'Have you had anything to eat at Widow Hodge's?' she asked. He shook his head. 'No.'

'Then we'll share this chicken. There's bread I baked today in the larder, and some cheese and a few pickles.'

'All right – but you can also have the chicken I was taking with me. I won't need it if I'm eating with you.'

When Gideon left the Treleggan rectory more than an hour later he was comfortably full and knew a great deal more about Alice and the unhappy life she had led. He had also told her far more of his own life than he was used to disclosing to anyone. They had talked to each other with increasing ease and he had left promising to return to see her again as soon as he was able.

When he did, Gideon decided he should return by night once more. Not because *he* feared anything from the young men of Treleggan, but to avert any unpleasantness an overt visit might cause for Alice.

6

I

Gideon paid evening visits to Alice at the rectory on two more occasions over the next few weeks. Then, learning it would be her twenty-first birthday the following Saturday, he made the surprise suggestion that she should spend the whole of that day with him, away from the rectory and Treleggan.

He explained that the Friday before her birthday would be settlement day for his gang. They would be paid their bonuses for completing the cutting on which they had been working ahead of time, and settle their debts with the 'Tommy-men'. Afterwards, it was customary for the navvies to spend the remainder of the weekend drinking away much of their hard-earned money.

Despite their weekend binge – known as a 'randy' in the parlance of the navvy – every man of them would report for work on the site of the new cutting, half a mile along the line, early on Monday morning.

When Alice asked where they would go for a whole day, Gideon said it would be a surprise – but suggested she should dress as if she were going to a Sunday church service.

Alice was not at all certain about leaving the rectory for a whole day to go out with someone she still felt she knew little about. Not until Gideon promised he would return her soon after eight in the evening did she eventually agree. Nevertheless, she had her doubts when he suggested coming to the house for her at eight o'clock on the morning of the planned excursion.

'Wouldn't it be better if we met somewhere outside the village? We've been very sensible so far, why stir the villagers up unnecessarily?'

'I've thought about that, Alice,' Gideon replied seriously, 'thought about it very carefully. I feel it's time we brought our meetings out into the open. It would cause more of a scandal if they learned I'd been visiting you secretly, after dark. Besides, it's your birthday.' Aware that Alice was not entirely convinced, he added, 'How can we really get to know each other if we only meet after dark in the kitchen of a rectory, constantly afraid someone is going to find out about us?'

Thankful that he had no way of knowing how much his words thrilled her, Alice asked, 'Do you want to get to know me better?'

'Of course I do.'

He was obviously sincere and Alice said, 'All right then, I'll come with you on Saturday.'

That night, alone in her attic room in the Treleggan rectory, Alice fell asleep speculating about Gideon's intentions towards her.

On the Saturday morning of her birthday, Alice was ready long before eight o'clock and becoming increasingly excited. She had arranged for Ivy Deeble, the housekeeper, to come to the rectory at nine o'clock and remain until dusk. But apart from providing company for Digger there would be little for the housekeeper to do. Alice had worked until late the previous evening to ensure that everything in the rectory was spotless.

Not that there was anyone to make the house untidy now. Only the kitchen, scullery and Alice's attic room were in regular use. A daily attack with a feather duster and a weekly window cleaning exercise were sufficient for the remainder of the large house.

Waiting in the front hall, Alice wore her best gingham dress and a light shoulder cape that had been a present from the wife of a curate who had lodged at the rectory for a few weeks. She was excited, yet apprehensive about the day ahead. She also had an irrational but deep-seated fear that Gideon would fail to keep their assignation.

By the time the deep tones of the massive hall clock struck the hour, Alice had already convinced herself that he was not coming for her. By five minutes past eight she was so agitated she was wondering whether to go out to the rectory gateway to see if she could see him on the road.

Then, at the height of her despair, Gideon arrived. He was not on foot, as she had been expecting, but driving a pony and trap.

He had hardly brought the pony to a halt when Alice emerged from the rectory and Gideon looked at her approvingly. 'You look beautiful, Alice. I'm glad I hired a pony and trap. Nothing else would have done – except, perhaps, a carriage!'

Trying hard not to let him see how much his compliment pleased her, Alice said, 'Well, here I am – but I still don't know where you're taking me.'

'That can remain a secret for a while longer,' Gideon replied mysteriously. 'I don't think you'll be disappointed – but let me help you into the trap. We mustn't be late.'

After commenting that she wished she knew what it was they must not be late for, Alice allowed Gideon to settle her comfortably on to the padded seat. Seating himself beside her, Gideon guided the pony out of the rectory gates before setting it off at a brisk trot. As they made their way between the houses, Alice was aware they were attracting a great deal of interest from the inhabitants.

Riding in a pony trap was a new experience for Alice, which was only slightly marred when they passed Billy Stanbeare walking towards the village from his father's farm. It was the first time she had seen the bully since his fight with Gideon and she was taken aback by the disfigured nose.

Observing her expression, Gideon said, 'Don't allow the sight of Billy Stanbeare to spoil your day, Alice. He won't make your life miserable any more.'

Alice was less confident than Gideon that Billy would cease his harassment of her, but she kept her thoughts to herself and even managed a smile.

'Have you ever driven a pony and trap, Alice?' Gideon asked.

She shook her head. 'It's the first time I've ever *been* in one.'

'Then today is going to be a day of many firsts for you. Here, take the reins and drive for a while. You'll enjoy it.'

Although Alice protested she could not possibly manage the trotting pony, she took the reins and found she *did* enjoy controlling the animal – even when they overtook a slow-moving hay-wagon on the narrow lane and she squealed with momentary fright.

She was enjoying Gideon's company too and as the sun climbed higher into the sky there seemed every prospect of a fine day ahead.

It was not long before the road began to descend a steep hill into a valley, through which the main road between Bodmin and Liskeard ran alongside the river Fowey. Beyond river and road was a section of the newly constructed railway line.

At the foot of the hill was an inn, its garden bordering the river. Gideon had retaken control of the reins a short time before and now he turned the pony in to the inn yard and brought it to a halt. After helping Alice to the ground, he passed the pony and trap into the keeping of a liveryman and it was apparent to her that they had been expected.

Alice was more mystified than ever now, but she decided to put herself entirely in Gideon's hands and enjoy the day in the same way she had enjoyed the journey from Treleggan.

Despite her decision, she was surprised when, instead of taking her inside the inn, as she had expected, Gideon led her out of the yard and away from the hostelry.

They passed over the bridge, crossed the road and then began walking along a narrow path that led up the far side of the valley. Here Gideon took her arm to support her, should she lose her footing.

A few minutes later they passed through a small gate and on to the railway line. There was no one else in sight, but a small wooden platform had been constructed alongside the line, with four steps leading up to it.

Instructing Alice to mount the steps, Gideon gave her a hint of what they would be doing. 'Have you ever been on a train, Alice?'

'I've never even *seen* a train,' she replied. 'Why . . . we're not going on a train today?' She sounded scared. 'I'm not sure I'm ready for that!'

'You'll be fine, Alice. People all over England are travelling on trains now, going to places they would never have been able to visit had it not been for the railway. Mind you, not many will have had a special platform built for them. Tell me, what's the farthest you've ever been from Treleggan?'

It was a question he bitterly regretted asking when Alice replied, 'Ma and me travelled around Cornwall quite a bit. Bodmin, Callington . . . and we spent a long time in the work-house at Liskeard . . .'

'Those are unhappy memories, Alice. Today will be quite different. It's going to be a special day for you – as a twenty-first birthday should be.'

Deciding it was time he told her what he had planned for the day, Gideon said, 'The train is coming from Liskeard with passengers on an excursion to a spot near Charlestown, on the coast, where there'll be carriages and wagons to take us to Charlestown harbour – we'll be travelling in a carriage. Waiting just offshore will be a paddle-steamer on an excursion from Plymouth. It will take us to Falmouth where we'll have time for something to eat and a look at the shops before boarding a small steamer to travel upriver to Truro. We'll have time to look around more shops there before catching a train to bring us back here. Then you'll go home again in the pony trap. So in one day you'll have been on two trains, two boats, visited two towns, twice ridden in a pony trap, and had a meal at the finest inn to be found in Falmouth! Does that sound a special enough day for your twenty-first birthday?'

Finding it difficult to take in all that he had arranged for her, Alice said, 'But . . . but *why*, Gideon? Why have you gone to so much trouble for me?'

'You've had a rotten time lately, Alice – in fact, from what I've heard you've had an unhappy time for most of your life. I want to give you a birthday you'll remember as a particularly happy day.'

Alice was deeply touched by his words, but he had still not explained why he wanted her to have a happy day. She would have persisted with her questioning, but at that moment they both heard the hoarse shriek of a steam whistle. When Alice looked along the railway line to the east of where they were standing she saw her first train. Coming round a bend farther along the line, it was heading towards them, smoke belching from a tall smoke-stack.

Pulling an impressive line of small carriages, the locomotive was much larger than she had imagined it would be. The smoke-stack stood at least as tall as three men.

The train was slowing now, and soon the engine passed Alice and Gideon with a frightening hiss of escaping steam and a cheery wave from the driver. When it came to a halt, Gideon opened a carriage door and helped Alice inside. Then, waving to the driver, he climbed in and closed the door behind them.

There were three other occupants of the carriage: a man of about eighty, a woman perhaps half that age and a young girl who looked as though she was about sixteen. Taking a window seat opposite the young girl, Alice observed a similarity in the features of the trio and realised they must be three generations of the same family.

Gideon apologised for their sudden and unexpected arrival, to which the old man said ungraciously, 'I hope the driver does not intend stopping for everyone he sees standing beside the line!'

'It was a prearranged halt,' Gideon explained, in an effort to placate the old man. 'There will be only two more between here and Charlestown.'

Further conversation was rendered impossible when the train set off once more, its initial progress marked by a series of neck-jerking jolts. The uncomfortable movement ceased as the train gathered speed and Alice stared wide-eyed as trees rushed past the windows.

Turning to Gideon, she commented, 'We seem to be going very fast.'

'It's the quickest way to travel anywhere,' Gideon agreed enthusiastically. 'Probably the fastest means of travel there will ever be. Engineers are building a railway bridge across the River Tamar right now. When it's completed it will be possible to enjoy breakfast in Liskeard, then catch a train that will have you in London in time for an early supper.'

The old man had been following the conversation between Gideon and Alice, but when he spoke it was to his daughter. 'I don't know why there is such an obsession today with speed. Travelling is intended to be leisurely, to give a man time to think before he gets to wherever he's going. It isn't natural for man to move so fast. Had that been what God intended, we would all have been born with wings.'

'Father! It is extremely rude to make comments about a conversation in which you were not included.' Having admonished the old man, the woman said to Gideon, 'Please excuse my father's rudeness. He is eighty-one and I am afraid that a lifetime spent at sea, in command of ships, did nothing to improve his manners.'

'And too many years without the discipline of a father did nothing to teach you respect for your elders,' retorted the old man. 'No one has ever needed to apologise for the behaviour of Captain Gilbert, and I won't have my own daughter doing it now.'

'There is no need for an apology,' Gideon said hastily, in a bid to avert an argument. 'Travel by train is something very new that we will all need to become used to. However, when we reach Charlestown we will be travelling in a manner far more to your liking. I believe the ship taking us to Falmouth is one of the very latest and largest paddle-steamers. Are you acquainted with such vessels, sir?'

Shifting the conversation to ships sufficed to put the ex-sea captain in an immediate good humour. He launched into details of the ships he had commanded, while his daughter sat staring fixedly ahead of her, stony-faced, and the granddaughter gave Alice a relieved smile.

By the time they reached the next station along the line, Alice and Gideon had learned that Captain Gilbert had commanded many large sea-going vessels, that his daughter, Charlotte Massey, was the widow of an army officer, and the grand-daughter's name was Deirdre. Then the compartment was invaded by a family which included four lively and excited young boys, and the captain's monologue came to an end.

II

There was no station at the tiny port of Charlestown but impro-vised steps had been provided to allow passengers to alight. However, there were far more compartments to the carriages than steps. As first-class passengers were given priority, many of the men occupying other carriages jumped from their compartments and helped women and children to the ground.

There was similar discrimination when it came to transport from the railway line to the harbour. Carriages were on hand for first-class passengers. Others either made the journey in straw-strewn open wagons, or walked.

Gideon had ensured that Alice would ride in a carriage. As a result, they were among the first of the passengers to reach the harbour. Here they boarded a small steam launch that carried them to a large and impressive paddle-steamer waiting a short distance from the shore.

Passengers from the train were embarked on the ship in a surprisingly short space of time, and before Alice had settled herself in her new surroundings the huge paddle-wheels began to churn the water to froth on either side of the vessel and it got under way.

It was the first time Alice had been on board any form of boat, but she found it far less frightening than the train and soon decided that the Cornish coast was truly beautiful with its hidden coves and fishing villages.

The sea journey ended at Falmouth where the ship berthed alongside a jetty and the passengers were able to walk ashore over a gangway.

Gideon was familiar with the bustling port, explaining to Alice that he had worked on a railway that was intended to connect Falmouth with Truro. Unfortunately the contractor had run out of money and the work was brought to an end before the line was completed.

Because of his knowledge he was able to take her straight to an inn overlooking the harbour, promising that when the meal was over they would have time to look round the town's many shops.

The lunch surpassed anything Alice had ever experienced before and she told Gideon so.

Pleased that the meal was a success, Gideon said, 'So you are enjoying your birthday treat?'

Beaming at him, Alice said, 'I'm having a wonderful day, Gideon . . . thank you. Thank you *very* much.'

Her obviously genuine pleasure delighted Gideon. He had not known Alice for very long, yet it was long enough for him to realise his feelings for her went far deeper than anything he had ever experienced before. Because of this he had gone to a great deal of trouble to ensure the day would be one she would always remember.

Gideon's gang of navvies were aware that Alice was someone special for him. As a result he had undergone a great deal of mainly good-humoured chaffing from the rough-and-ready men, but they had been happy to build the small platform in order that she might board the train in safety.

*　　*　　*

52

When Gideon and Alice left the inn, Alice, in particular, was looking forward to browsing round the shops. However, they had not gone very far from the inn before they were accosted by Charlotte and Deirdre Massey. The older woman, in particular, was extremely agitated.

Obviously relieved to see someone she knew, albeit only casually, Charlotte said, 'My father . . . have you seen him anywhere? He has gone missing!'

Gideon shook his head. 'We haven't seen him since he was on the ship, with you. How have you managed to lose him?'

Her glance darting here, there and everywhere as she spoke, Charlotte explained, 'Deirdre and I wanted to look at the shops. We felt it would be too tiring for him, so we left him seated on the quay, not far from where we landed, in order that he might look at the ships. When we returned for him, he had gone!'

'How long were you away?' Gideon asked. It was now well over an hour since the paddle-steamer had landed them at Falmouth.

'No more than twenty minutes or so . . .' Her daughter gave her a sharp look and the woman amended her estimate. 'Well, perhaps forty-five minutes – it certainly wasn't a full hour.'

'Perhaps he became bored,' Gideon suggested. 'Let's go back to where you left him and see if there's anything to be learned there.'

Alice was disappointed not to be able to spend the time browsing in the shops, but she realised this was something of an emergency.

The small group walked a short distance along the street before turning into a narrow alleyway. When they emerged on the quay, Deirdre pointed to a wooden bench seat, saying, 'That's where we left him.' Turning wide eyes upon Gideon, she added, 'You don't think he might have fallen into the water?'

Gideon shook his head. 'There are a lot of people about. Someone would have seen it happen and there would still be a lot of excitement going on.' Looking about him, he espied a tavern on the town side of the quay and asked Charlotte, 'Is your father a drinking man?'

'Certainly not!' The woman drew herself up indignantly, before drooping once more and adding, reluctantly, 'Of course, he has an occasional glass of wine with his dinner.'

'And whisky – and port – and brandy,' Deirdre declared. 'You are always telling him he drinks far too much, Mother, you know you are.'

Breaking in on what he believed might become an acrimonious exchange, Gideon said, 'You stay here, with Alice. I'll go into the tavern and see if he's there.'

Leaving Charlotte protesting that her father would never drink in 'a common dockside tavern', Gideon gave Alice an apologetic glance, then hurried to the establishment to which she referred.

Entering the bar room, Gideon conceded that Charlotte's description might have been more accurate than he had at first thought. This was not a place that would be patronised by a respectable elderly gentleman. The clientele consisted of sailors from many nations, together with a number of women who earned their living gratifying the urges of men who had been at sea for many weeks.

Despite the unlikelihood that Captain Gilbert would be found here, Gideon's glance searched among the many customers without success. He was about to leave when the landlord of the tavern confronted him and said, 'Are you looking for anyone in particular, sir?'

'Yes. An old gentleman who's gone missing off the quay. His daughter's concerned for him, but I can see he's not here.'

'Would this gentleman be grey-haired, with heavy side-burns and a silk top hat – and talking of once being a sea captain?'

'You've seen him? He *was* in here?' Gideon's disbelief was apparent.

'I served him myself,' the landlord replied.

'I suppose you've no idea where he went when he left?' Gideon asked, without much hope of being given a lead to Captain Gilbert's present whereabouts. He added, 'He's here on an excursion with his daughter and granddaughter and is due to go upriver to Truro in less than half an hour.'

'I couldn't tell you *where* he went,' the landlord replied, 'but I can tell you who he went with. He left here with Rosie . . . Rosie Miller.' The landlord seemed highly amused.

'Left with . . . ?' Gideon looked round the room at the girls who were seated at the tables, sharing jokes and beers with the merchant seamen. 'You don't mean one of *these*?'

The landlord's amusement grew. 'That's exactly what I mean. He's a randy old codger, and no mistake – but I don't know how you'll explain the situation to his daughter.'

'I'll worry about that later,' Gideon said. 'What's important at the moment is finding him in time to catch the boat to Truro. Can you point out any of the girls who might know where this Rosie Miller lives?'

The landlord shook his head. 'Even if I could, you'd get nothing out of any of 'em.' After a moment's hesitation, he added, 'At this time of day you'll probably find Constable Julyan in the Custom House. He'll know where Rosie lives, he's arrested her twice this year already.'

Thanking the landlord, Gideon hurried from the tavern, uncertain what he would say to Captain Gilbert's daughter.

He found Charlotte in a highly agitated state once more, with Alice and Deirdre trying to calm her. When she saw Gideon, she pushed the others to one side and hurried to meet him.

'Have you learned anything? Has anyone in there seen what happened to him?'

Trying to soothe her, Gideon said, 'Yes, I think I know where he is – and there's nothing to worry about, he's all right, but I need to find the town constable to show me where the place is. You remain here with Alice and Deirdre, I'll not be long.'

'I am not remaining *anywhere*!' Charlotte declared adamantly. 'I'll come with you and give him a piece of my mind, going off and worrying me in such a manner!'

'I really do think it would be better for you to stay here.' Adding a lie he hoped she would believe, he said, 'He's gone off with some new-found friends to a rather insalubrious part of town. It's no place for you, or for your daughter.'

'I am coming with you.' Charlotte spoke in a tone of voice which allowed no further argument. 'Deirdre will come too. She is always complaining that I am far too hard on her grandfather. She can come along and see for herself what a foolish old man he can be when I am not around to keep an eye on him every minute of the day.'

Gideon thought he could understand why the old man had taken the opportunity to escape from his domineering daughter for an hour or so – but finding him was a matter of urgency if they were all to catch the boat to Truro.

He gave in without further argument and, with himself and Alice leading the others, they hurried to the Custom House.

When Gideon asked for the constable, he was taken to an inner office, while the women were left waiting in the reception area. He was introduced to Constable Julyan and hurriedly explained the situation, adding that the old sea captain's daughter was insisting upon accompanying them when they went to find her father – and explaining that she was an extremely positive woman.

The constable grinned. 'I hope she's also broad-minded. Rosie's no respecter of persons, whoever they may be. She has a vocabulary that's the envy of every sailor she meets with – and she meets with more than most!'

In spite of his words, Constable Julyan tried to persuade the three women to remain at the Custom House to await their return, but Charlotte was as stubborn as before. She was coming with them, and Deirdre was coming too.

Gideon would have preferred Alice to remain behind, but he had seen Deirdre give her such a pleading look that when she said she would come too he did not argue with her.

'Please yourself.' The constable shrugged. 'It will be a memorable experience, I've no doubt.'

Less than four minutes' walk from the dock area, Constable Julyan, walking ahead of the others, turned into a narrow, rubbish-strewn alleyway.

An expression of extreme distaste on her face, Charlotte followed him, hitching her skirts clear of the filth on the ground. The other women did the same.

There were no numbers on the doors which opened on to the alleyway, but Constable Julyan was spared the problem of deciding which doorway led to the room occupied by Rosie Miller.

Coming towards the small group was Captain Gilbert, his arm linked through that of a woman who, as they watched, laughed and leaned her head on his shoulder.

The old man gave a start of surprise when he recognised his daughter and granddaughter among the group approaching him. Then his shoulders stiffened, he straightened up to his full height and adopted an air of exaggerated dignity. Addressing his daughter, he demanded, 'What are you doing here? You should be at the pier, catching a boat to Truro.'

'*I* should be . . . ?' Charlotte bristled with indignation. 'You go missing, have half the town looking for you, then tell *me* what I should be doing . . . !'

'You've reported me missing?' Captain Gilbert appeared genuinely astonished. 'Why? I'm a grown man, I can do what I want to do. I am not a small child to be put on a seat and told to stay there out of mischief until you deign to return and order me to do something else. Neither am I a piece of baggage, to be set down and picked up at your whim. I became tired of waiting for you and went for a drink. In the tavern I met up with Rosie and she took me to her home. We discovered we have much in common . . .'

'"Common" is the right word,' Charlotte snapped at him. 'She should be arrested. As for you . . . I am ashamed of you, Father. You have disgraced yourself in front of Deirdre – and in front of strangers, too.'

'You take no notice of her,' Rosie told Captain Gilbert. 'You've got nothing to be ashamed of, nothing at all.' Turning her attention to Charlotte, she said, 'There's more life in him than I've known in men half his age. You should be proud of him.' After a last, scornful glare at the suddenly speechless

Charlotte, Rosie spoke to Constable Julyan. 'Before you say anything, I've committed no offence. We'll both deny that any money's changed hands and I've told the captain he's welcome to come back and see me as often as he likes. You don't meet gentlemen like him very often these days.'

III

The time Alice and Gideon were able to spend wandering around the shops in Truro more than made up for not being able to see more of the ones in Falmouth. Alice bought herself a beautiful Indian shawl and when she expressed particular admiration for a silver, Celtic-style crucifix, Gideon purchased it for her as a birthday gift.

When Alice protested that he had already spent far too much money on her, Gideon replied that until he had met her there had been no one on whom he had ever *wanted* to spend his money.

His words added to the warm feeling that had been growing within Alice during their day together. As they walked around the Cornish town, her arm linked through his, she frequently reached up to touch the crucifix she wore on a chain about her neck. When she did this she would squeeze his arm in an expression of affection that delighted Gideon.

He was disappointed that the feeling of warm intimacy could not be continued on the train journey back to the improvised railway platform from which they had started the day's excursion. He had hoped they might have the compartment to themselves, but at the last minute three perspiring farmers, complete with wives and children, crammed into the restricted space. It came as a relief to Alice and Gideon when they eventually escaped from the overcrowded compartment and made their way down the hill to the Halfway Inn.

The arrival of the train had been witnessed by the liveryman and he had a pony harnessed to the trap and ready for them in the yard by the time they arrived at the inn. By now the sun had set beyond the hills, but the sky to the west still carried the red and orange colours of evening.

The lanterns fitted on either side of the pony cart had been lit by the liveryman, but they were not really required yet and as the pony set off up the hill Gideon asked Alice whether she would like to take the reins once again.

'No, Gideon, you drive, I want to just sit here and think about the lovely day I've had.'

Smiling at her, he said, 'I'm glad you've enjoyed it, Alice.'

'*Enjoyed*? It's been the most wonderful day I've ever known. I wish it was morning and that we could have the time all over again.'

Giving her a sidelong glance, he said, 'Even though old Captain Gilbert spoiled your chance to look round the Falmouth shops.'

When Alice laughed, Gideon thought it a delightful sound. 'He didn't spoil anything, not really. In fact I found what happened there highly amusing. I think Deirdre did too. It was only Charlotte who was so shocked by what her father had been up to. I wouldn't think she was a lot of fun for either her father or Deirdre to live with.'

She fell into a thoughtful silence before saying, 'You know, of all the people we've met up with today, the one I feel most sorry for is the girl Captain Gilbert went off with . . . Rosie.'

Gideon looked at Alice in surprise. 'You feel sorry for her? I would have thought most women would feel revulsion.'

Alice shook her head. 'No, something awful must have happened to make her take up such a horrid way of life. No one would *choose* to live that way. I know that Florence and Fanny, who work at Widow Hodge's kiddleywink, wouldn't do what they do if there was another way they could earn money to keep their family. It must be horrible.'

In truth, Alice was thinking of her own life as much as that of Rosie and the others. She had often wondered what would have become of her had Parson Markham not come into her life when he did. She was also well aware of the uncertainty with which she was now faced . . . but this was not the time for such thoughts.

In a determined effort to put them behind her, she once more reached up to touch the crucifix Gideon had bought for her and began talking about other incidents of the day. The excitement, tinged with fear, when she saw her first train; the voyage to Falmouth in the steamer; shopping in Truro and, most of all, Gideon's great kindness to her. Not until they came within sight of the lights of Treleggan did her happiness begin to seep away again.

Suddenly serious, she asked, 'Gideon, why did you spend so much time and trouble on arranging this day out for me?'

'Why . . . ?' Gideon suddenly felt tongue-tied. 'I . . . I did it because it's your birthday, a very special birthday.'

'And you've made it a very special day for me, but why did you *want* to?'

'Do I really need to explain that to you, Alice?'

Gideon had guided the pony into the rectory driveway now, but before Alice could reply to his question they both saw an indistinct figure run from the front door of the rectory and disappear along a path that led through trees to a small gate at the side of the large garden.

Bringing the pony to a halt, Gideon leaped down from the trap. Telling Alice to remain where she was, he set off at a run in pursuit of the trespasser.

It was a hopeless chase. By the time Gideon reached the gate the intruder had disappeared into the darkness. It would have been futile to search for him.

Returning to where he had left the pony, he found that Alice had not remained in the trap. The door to the rectory stood open and a lamp had been lit in the hall. As Gideon approached the house, Alice appeared in the doorway. Even in the poor light, he could see tears glistening on her cheeks.

Alarmed, he asked, 'What's the matter, Alice? Has someone broken in?'

She shook her head. 'I don't think that's what they were doing here . . . look.' Standing to one side, she gestured towards the door.

Hurrying forward, Gideon drew in his breath in an expression of anger. Painted upon the door in large, white letters was the word *WHORE*.

IV

Later that evening, Gideon and Alice were seated in the rectory kitchen, drinking tea and discussing the vandalising of the front door.

The paint used had been a limewash. As it had not had time to dry it was not too difficult to clean off with soap and warm water. Fortunately, Ivy the housekeeper had placed a kettle on the hob before locking the rectory and going home. Once the limewash had been thoroughly removed, Gideon used the remainder of the hot water to make a cup of tea for the badly shaken Alice.

When she was drinking her second cup, Gideon asked anxiously, 'Are you feeling any better now?'

Alice nodded. 'Yes, thank you – but who would have done such a thing, Gideon – and why? Why do the villagers hate me so much? I've never done anything bad to anyone in Treleggan. I know Ma made a mistake, but that was years ago and she paid dearly for it. She was the one who was hurt, not them.'

'I think you already know the answer to your question, Alice – and I'm partly to blame. Your ma's biggest mistake was to upset a man with some influence in the village. Now I've come along, an outsider like your father, to open up old wounds and stir things up again, this time for you. I'm afraid Henry Stanbeare isn't the forgiving kind – and you're an easy target.'

'But what's happened tonight won't be Henry Stanbeare's handiwork, surely?' Alice said.

'No, it's more likely to be the work of Billy. He's probably trying to get at me through you. If you remember, he saw us leaving the village together this morning.'

'It might not even be Billy,' Alice said unhappily. 'The villagers wouldn't need much persuading to turn on me, especially if it was what Henry Stanbeare expected of them. You saw as much of who did the painting as I did and it certainly wasn't big enough to be Billy.' Thoroughly dejected, she added, 'I know you've been very careful, Gideon, but you've been seen coming to the rectory. Word has gone around that I'm entertaining a "foreigner" here, after dark. I was told as much the other day when I bought my milk up at the farm.' Close to tears once more, she said, 'I just wish this hadn't happened tonight, after I've had such a lovely day.'

Gideon felt an overwhelming urge to take her in his arms and comfort her, but this was not the moment for such a gesture. She might believe he was taking advantage of the situation and had no more respect for her than had the villagers. Unless clear encouragement came from her, tonight was not the right occasion to take their relationship any further forward.

The realisation disappointed him deeply. Nevertheless, Gideon was determined that neither Billy Stanbeare nor any of the villagers would succeed in causing a permanent rift between Alice and himself.

'Coming here to see you at any other time than the late evening isn't easy for me, Alice, but would it be better if I only came to see you during the day, on a Sunday, perhaps?'

'It might,' Alice conceded, 'but I think they've already made up their minds about what we're doing. If they can't prove anything they'll make it up.'

Angry to see her so unhappy, Gideon said, 'This is no life for you, Alice. You shouldn't have to put up with people behaving this way towards you.'

'It's nothing new,' Alice replied despondently. 'They've always made it clear that I don't really belong in Treleggan. I've got by because I could shut them all out whenever I closed the rectory door – until tonight, that is.'

'No one should have to live in such a way,' Gideon persisted. 'Especially as you've never done anything to harm anyone.'

'I'll put up with it . . . at least, I will until the new rector arrives. What happens then depends upon what sort of man he is.'

'There's no need for you to leave your fate in his hands, Alice. You can leave the rectory whenever you like . . . tonight, if you wish.'

Looking at Gideon as though he had suddenly taken leave of his senses, Alice said, 'Leave the rectory? What do you suggest I do instead? Where can I go – to lead the sort of life led by Rosie?'

'No, Alice. I'm suggesting that you let me take you away from here and find somewhere for you to live . . . until we can get married.'

Alice looked at Gideon in utter disbelief. 'Get *married*! What are you saying, Gideon? We hardly know each other!'

'We have a lot to learn about each other,' Gideon admitted, 'but I already know enough to believe we could be very happy together.'

After remaining silent for some moments, Alice shook her head vigorously. 'You're trying to bring a dream to life, Gideon. You've not had an easy life yourself. Now you've met me, who you believe life has also treated badly. Yes, we do have a lot in common, but that's not enough to make a success of marriage.'

'There's a lot more to it than that, Alice. I've never felt for anyone the way I do about you. I find myself thinking about you when I have time on my hands, and also during the day when I'm working. I want to be with you and have you with me for always.'

Gideon was making the first declaration of love that Alice had ever received. Her instincts were to respond in a similar manner, but the oft-expressed admonitions of her mother, coupled with the lessons she had taken from life itself, prompted caution.

'I'm flattered, Gideon . . . very flattered indeed, but it's far too soon to be talking of marriage between us.'

Gideon felt dejected at her refusal, and it showed. 'I'm sorry, Alice. I didn't want to hurry things in this way, but after what's happened tonight I wanted you to know how I felt. To show you that it isn't necessary for you to put up with things the way they are.'

'Thank you, Gideon. Thank you for letting me know I'm not entirely on my own. It means a great deal to me, it really does – especially now. But I would need to know you a lot better than I do now before I thought about marriage. I'd want to be certain that if we did marry it would be for the right reasons and not because you felt sorry for me – or because it was an act of desperation on my part.'

Gideon said nothing for so long that Alice feared she had deeply offended him. She was wondering how she might set it right when he said, 'You're probably right, Alice. *I* have no doubts at all about wanting to marry you – and it has nothing to do with feeling sorry for you. But I wouldn't want you to agree to marry me until you were quite certain you felt the same way.'

'Thank you, Gideon.' She was greatly relieved. 'I hoped you would understand.' Reaching a sudden decision that surprised her, she said, 'I . . . I really do think I feel the same way about you, but I need to be absolutely sure. Does that make you any happier?'

As the import of her unexpected statement sank in, Gideon could hardly contain his joy. 'Yes, Alice. Yes, it makes me much, much happier.'

'I'm glad, because until we got back to the rectory tonight I'd had the happiest day of my life. I'll never forget that. Now I think you had better go and leave me to tidy things up in here, ready for tomorrow.'

Suddenly remembering what had happened, Gideon asked, 'Are you quite certain you're going to be all right here by yourself? You don't want me to stay around for a while . . . outside, if you would prefer?'

Alice smiled. 'I don't think that will be necessary. Painting a word on the door is one thing, doing me actual harm is something very different. When you go I'll lock the door and let Digger guard me.'

7

I

The day after the excursion to Falmouth and Truro was a Sunday and Gideon would have liked to pay another call on Alice. However, in view of the trouble that appeared to have been stirred up in Treleggan, they had decided he should defer his visit until the following week.

Instead, he left the camp and walked along the line to the site of the section on which he and his gang would be working for the next few weeks. The work involved a number of cuttings and embankments and he wanted to satisfy himself they would be able to complete it well on schedule.

He was standing at the site of one of the cuttings, making mental calculations, when he looked up to see a horse and rider heading towards him, picking their way through the chaos that existed around the railhead.

As they drew closer he recognised the rider as Robert Petrie, the man who held the contract for the whole of the new line between Penzance and the proposed railway bridge across the River Tamar.

In common with Gideon, Petrie had begun his working life as a navvy. A deeply religious man, he also possessed a burning ambition. It was this which had been responsible for his rise from navvy to ganger and then sub-contractor before becoming a successful and wealthy contractor, respected by every railway engineer of note in the country. Petrie demanded the very best from every man who worked for him, but he paid them fair wages in return.

Gideon both liked and respected the contractor. Petrie, in his turn, recognised in Gideon the drive that had motivated him more than twenty years before.

Reaching the spot where Gideon stood, Petrie reined in his horse and greeted Gideon by name before dismounting and shaking the ganger's hand.

'I thought I would find you here. Are you anticipating problems?'

Gideon shook his head. 'No, we'll have the job completed well on time. I came up here to reassure myself. I don't want to be greedy, but I do want to make a profit.'

Nodding his approval, Petrie said, 'That's the only way to do business, Gideon. When I was a ganger I always tried to do the same, but I had a goal in mind . . . do you?'

'I suspect it's the same one you had in those early days,' Gideon replied. 'I want to make enough money to enable me to sub-contract for work that will involve a number of gangs until one day I'll be able to tender for a whole line.'

'It could mean we'll be bidding against each other at some time in the future,' the contractor commented, without any apparent rancour.

'It could,' Gideon agreed, 'but I doubt it very much. I'll be bidding for branch lines at first. You're a main line contractor – and when they're all completed in this country the rest of the world will be crying out for contractors with your expertise.'

'I can see you've thought things out well, Gideon, but, as it happens, I'll probably not be contracting for very much longer. I've been asked to stand for Parliament at the next election.'

Gideon was taken aback that the contractor should reveal such plans to him, but there were more surprises to come.

'Before I make up my mind I and a few other contractors have been asked by the Prime Minister to pool our resources and build a particular stretch of line. It will extend for only a short distance, but is very, very important. That's why I've come to speak to you today, Gideon. I would like your help.'

'You want *me* to help *you*?' Gideon could not hide his astonishment. 'What's so important about this particular line? Where is it?'

'In the Crimea, where our army is fighting against the Russians. We're in serious trouble there, Gideon. From what I've learned, our generals are spending more time fighting each other than battling with the enemy. But, as well as holding a line against the Russians, they're investing the town of Sebastopol. Unfortunately, the only supply port available to them is miles away. It's far too late for them to find another harbour now and with winter setting in it's feared the very poor roads – where they exist – will become impassable, putting our soldiers in very real danger from starvation. As it is there's so much sickness among them that if something isn't done quickly we'll have no army left by the end of the winter.'

Gideon was not even certain where the Crimea was and he said so.

Petrie gave a brief smile. 'The first thing I did after speaking to the Prime Minister was to go off and buy myself an atlas. The Crimea is on the Black Sea – that's at the eastern end of the Mediterranean – and about three hundred and fifty miles beyond Constantinople. It will be a challenge merely to get materials there, but I've agreed to help build a railway between the depot, at Balaklava, and the siege lines around Sebastopol.'

Momentarily distracted by his horse, which was becoming restless at standing on a barren patch of ground when there was grass only a short distance away, Petrie loosened the reins to allow the animal to graze before turning back to Gideon.

'It's been decided I should go out there to see exactly what's involved, Gideon. I'd like you to come with me.'

Aware that Petrie was quite capable of assessing such a situation without help from him – or from anyone else – Gideon was taken by surprise. 'Why me? What can I possibly do that you can't?'

'You're ganger of the best navvies I've got working on this line. I want to take them all out there when we're ready to work, together with a couple of hundred like them. Before then I'd like you to look at what's to be done and assure me that the men are capable of completing the work in the time I tell the Prime Minister it will take. If you agree I'll want you to choose the navvies and be in overall charge of them. Have the line completed quicker than anything you've ever achieved before and you'll be paid accordingly. As far as the men are concerned I'll pay eight shillings a day and all found.'

The pay offered by Petrie was double the going rate for a navvy. Gideon knew he would have no difficulty in persuading his gang to accept, or in finding another two hundred good men. Such a contract should also probably provide him with sufficient money to become a sub-contractor. It would certainly increase his standing in the world of railway building.

'All right, I'll take it on and come with you,' Gideon replied. 'When would we be going?'

Robert Petrie smiled. 'I'm leaving for London this afternoon, Gideon. I'd like you to be with me. We're booked on a boat to France tomorrow.'

II

Gideon was relieved that 'Sailor' Smith was back in the navvies' camp and more or less recovered from his previous night's carousing. He would need to take over the navvying gang during Gideon's absence.

While he hastily packed personal items that were likely to be required on the unexpected journey to the Crimea, Gideon told Sailor what Robert Petrie had said to him.

'It will be enough to tell the men merely that I've gone off with Petrie to look at a line he's been asked to build and don't know when I'll be back. Petrie says we should keep the plans for the railway to ourselves for the time being.'

'You need have no fear of me saying anything, Gideon, but you say that Petrie wants us all to go to the Crimea to build this railway when everything is ready?'

'That's right, our gang and another two hundred good men. He's prepared to pay double the usual rate and take care of all expenses.'

'We'll earn every penny of it, Gideon. I visited that part of the world with the Navy some years ago. More than half the men on board fell sick just from eating the local food.'

'We'll take that into account when the time comes, Sailor. Is there anything else you need from me before I go?'

Sailor shook his head. 'You've got everything pretty well worked out for our next job, but don't be away too long. I can manage the men all right, but I haven't got your head for business.'

'You'll manage,' Gideon said confidently. 'There's one other thing I'd like you to do for me – and it's important. Call in at the Treleggan rectory and tell Alice I've had to go away with the contractor and will be gone for a few weeks. Tell her I'll write to her.'

'I'll do that.' Gideon had told Sailor about the painting of the rectory door and the navvy added, 'Should I go in daylight, or leave it until after dark?'

Gideon hesitated. If the villagers saw another navvy going to visit Alice it was likely to set tongues wagging again and cause more unpleasantness. 'It would probably be best if you went after dark – but make certain you're not seen or it will make things even worse for her. Avoid the village and cut through the churchyard to the side gate of the rectory. That way no one should see you. But there's no hurry about it. She's not expecting me until next Sunday.'

On the overnight train from Plymouth to London with Robert Petrie, Gideon learned a little more of what was likely to be involved in building a railway in the Crimea – and much about the contractor that he had not previously known.

Petrie too had started life with a good education and a future that seemed assured, but then his father went bankrupt and committed suicide as a result. Unable to cope with her grief, his mother had a breakdown in health and died only months later.

Left to his own devices and virtually penniless, Petrie's career followed the same path as had Gideon's, from navvy to ganger. When he became a sub-contractor, Petrie re-established contact with an uncle who had made his money investing in coal mines in the north of England and was able to persuade him to finance a bid for a railway contract.

The bid proved successful, the construction work had been completed ahead of schedule and Petrie found he was in demand as a railway contractor.

'I'm afraid I have no rich relatives in my background,' Gideon said ruefully, after hearing the contractor's story. 'And too many contractors have lost their money for any bank to consider offering a loan in order for me to tender for a railroad contract.'

'If we are successful with the line in the Crimea the banks will take a much more favourable view of your prospects,' Petrie said optimistically. He added, 'If they don't . . . well, I might decide to finance you myself, to retain an interest in the railway construction business while I go about my parliamentary duties!'

The contractor's words gave Gideon much food for thought. It could mean the realisation of his ambitions far sooner than he had anticipated. It also meant that he would have far more to offer Alice when he next proposed marriage to her. His imagination kept him awake when the other man was dozing in his seat in the otherwise empty train compartment.

The day Gideon arrived in London with Petrie proved every bit as hectic as the previous one. It had been arranged that the contractor would have a brief meeting with a general at the War Office, but three generals turned up, together with a number of less senior officers, and the meeting lasted all day.

Gideon was present with Petrie and it became increasingly apparent to him that none of the senior army officers had the slightest idea of the complexity of laying down a railroad, the immense amount of material needed – even what it would be capable of carrying when completed.

When the meeting eventually came to an end, the two railway builders emerged into the slightly foggy London street and Gideon was moved to say, 'I hope for the sake of our soldiers in the Crimea that their generals have more idea of what they're supposed to be doing than those we've just left.'

'I believe they're even worse than this lot,' Petrie replied grimly. 'If they were any good they wouldn't need *us* out there. Now, I'm going to have to make new travel arrangements. It's probably not such a bad thing. We'll have time to go shopping for some suitable clothes and luggage for you to take with you.'

On Tuesday the two men crossed the English Channel. Travelling from the port of Boulogne to Paris, Gideon and Robert Petrie took a keen professional interest in the French railways. They were both efficient and comfortable.

The journey was broken at Paris and here they were joined by Ranald MacAllen, a Scots railway engineer and surveyor in his thirties. He had been working on railway construction in France. A brilliant engineer, he had worked with Petrie before and had agreed to join him in the Crimean project. It would be his task to find a practical route for the railway, deciding where cuttings, embankments, tunnels and viaducts would prove necessary.

Gideon took to the engineer immediately. He had a quiet, gentle sense of humour and a manner that inspired confidence in his already proven ability.

After a pleasant evening in Paris, the three men boarded another train to Lyon, en route to the Mediterranean port of Marseille. The railway line had not been fully completed all the way to the southern coast and for a hundred kilometres, from

Lyon to the town of Valence, they would travel by boat down the River Rhône, and it was here that disaster struck.

As the trio waited at Lyon to board the boat, an arriving river launch sounded off its steam whistle. It was a strangled, shrieking sound that startled everyone within hearing and terrified a young horse standing in the shafts of a light cart at the water's edge.

The horse flung its head up in sudden fright, then bolted, heading straight for those about to embark on the boat to Valence.

Petrie and Ranald had their backs to the horse, watching Gideon who was a short distance away ensuring that luggage belonging to the three Englishmen was placed safely on board the river steamer.

Looking up, Gideon saw the horse bolting towards them, realised the danger the two men were in, and shouted a warning.

Ranald was able to jump out of the path of the terrified animal, but Petrie turned too late. The horse missed him by a hairsbreadth – but the cart did not. Petrie took a blow that sent him flying backwards over a guard rail and down a number of stone steps to the landing stage where the steamer was berthed.

Gideon too was bowled over by the runaway, but he fell back into a crowd of French men, women and children who were bidding farewell to family members. They all fell down in a tangled heap, but no one suffered more than very minor grazes and bruises.

Robert Petrie was not so fortunate. He remained lying on the ground, seemingly unable to rise, his face contorted with pain.

The scene at the riverside was one of pandemonium. The runaway horse had disappeared into the distance leaving carnage in its wake. Petrie was one of five injured and it was fortunate no one had been killed.

Gideon and Ranald tried to move Petrie to a more comfortable spot than the damp, cold stones of the riverside landing, but had to stop when he cried out in pain.

Looking around him, Gideon saw that, despite the drama that had just taken place, some of the passengers had recommenced embarking. 'I'll make certain our baggage doesn't go on board,' he said to Ranald. 'We won't be leaving Lyon today.'

By the time the steamer cast off from its berth a doctor had put in an appearance and some semblance of order had been restored to the scene of the accident. Arrangements were made for transport to take the more seriously injured to hospital, Petrie among them. When Gideon asked the doctor how serious were the contractor's injuries, the doctor shrugged non-committally, saying, 'Serious? I do not know. He has broken a bone here . . .' he pointed to the left collarbone, '. . . and here too.' This time he pointed to his ribs, adding, 'Elsewhere too, perhaps. I do not know.'

It was quite apparent that Petrie would not be fit enough to travel for some time and Ranald MacAllen and Gideon were at a loss as to what they should do about their journey to the Crimea and the railway that Petrie had proposed building there.

It was Petrie himself who supplied the answer when they called to see him that same evening. He had been told he must remain in the hospital for at least a week and further travel would be out of the question for at least another couple of weeks after that – more if he was contemplating the rigours of a sea voyage, with the inherent risk of falling and damaging bones that had not completely healed.

'So be it,' said Ranald MacAllen. 'We'll just have to wait in Lyon until you're fully fit.'

Petrie shook his head so vigorously he hurt his broken collar-bone and winced with pain. Gritting his teeth, he said, 'No, Ranald, this is one contract that can't be delayed. It's far too important. You and Gideon must go on without me and do whatever needs to be done. I'll follow as soon as I am able and take the work forward. In the meantime, I have letters in my luggage from the Secretary for War, the Commander-in-Chief of the army, and from the Prime Minister himself, requesting anyone in a position to help to do all they can for me – or for anyone acting as my agent. I'll write letters appointing both of you as my agents and confirming that you are acting on my behalf. Until I am well enough to join you in the Crimea you will both assume the considerable authority the Prime Minister has seen fit to bestow upon me.'

Gideon looked upon the responsibility he and Ranald were being asked to shoulder with considerable trepidation. He felt he should remind Petrie that he was only a ganger, inexperienced in the skills necessary to contract for a whole new railway system. But he remained silent. He did not want to add to the contractor's worries.

Aware of his doubts, Petrie said, 'Don't let the thought of the contract overawe you, Gideon. You can do it if you put your mind to what's entailed – and you'll have Ranald to help you. He's worked with me before and knows how I go about things. By the time we come to start work I'll be there to go over your figures, although I'm confident I'll need to change very little. The important thing right now is to set the whole project in motion as quickly as possible.'

After pausing to allow a spasm of pain to pass away, Petrie continued with his instructions, but eventually his distress became increasingly apparent and his breathing so laboured that Gideon and Ranald felt they should leave him to rest.

It would not be possible for the two men to travel on to Marseille that day. They found a small hotel in Lyon before going in search of a telegraph office from where Ranald sent a message to Lord Cowley, the British ambassador in Paris, giving him details of Petrie's accident.

He ended the message by assuring the ambassador that the 'mission' on which Petrie had been despatched would be carried out.

The two men visited the contractor in hospital the next morning before continuing their journey, but a French surgeon had administered anaesthetic before setting his broken collarbone and relieving the pressure on his lungs caused by broken ribs. As a result, Petrie was too drowsy to hold a lucid conversation. Ranald left a note for when he was more sensible, promising that his two 'agents' would ensure his contract was fulfilled.

This done, Gideon and Ranald set off on the next leg of their journey to the war-torn Crimea.

8

I

It was not until the Saturday after Gideon's unexpected departure from Cornwall that Sailor Smith walked to Treleggan. The week's work ended, his intention was to deliver Gideon's message to Alice. A number of Gideon's gang of navvies went with him as far as the high moor. They were going on to Widow Hodge's kiddleywink, where Sailor would join them later.

Night had just fallen when he reached the village, guided across the moor by the squat tower of the church, outlined against a cloud-scattered sky. Reaching the village, he made his way through the churchyard to the side gate that led into the rectory garden.

Once inside, he halted among the tall elm trees, wondering whether he should go to the front door of the rectory, or find his way to the kitchen entrance.

While he was trying to make up his mind he heard a faint squeak from the hinges of the gate he had just closed behind him.

Sailor moved off the path, seeking the shadows of the trees. Moments later a small figure passed by, heading for the front of the rectory. Leaving his hiding-place, Sailor followed, treading surprisingly lightly for such a large man.

There was sufficient light for him to see that the small figure ahead of him was carrying something. The thought had passed through his mind that it might be Alice, perhaps returning to the rectory after paying a visit to a friend, but he dismissed the idea almost immediately. He was aware from his conversations with Gideon that the rectory servant had no friends in the village.

When the unknown visitor passed from the shadow of the trees Sailor could see it was a boy, probably aged about twelve or thirteen.

The lad stopped at the rectory door but did not knock. Instead, he looked about him furtively. Sailor froze, knowing he would not be seen among the trees if he remained still.

He could see now that what the boy was carrying was a bucket. When he set it down in front of the door he produced a brush he had been carrying in a pocket. Sailor realised he was looking at the vandal who had daubed the limewash on the rectory door exactly a week before.

The navvy acted immediately – and quietly. He crept up behind the boy and seized him when the limewash-loaded brush was poised to paint another crude message for Alice. Pulling him away from the door, Sailor grasped his collar and lifted him clear of the ground.

The boy struggled violently, but Sailor merely twisted the collar even more savagely until the lad began to choke.

'Put me down! I wasn't doing nothing – honest!' he croaked.

'That's only because I came along and stopped you painting on the door and upsetting a young lady who's probably never done you any harm. What's your name, boy?'

When he received no reply, Sailor repeated his request, at the same time shaking the boy so vigorously he felt the rough-spun cloth of the shirt tear beneath his hand.

'Tim . . . Tim Yates. Stop shaking me. You've ripped my shirt. I'm going to be in trouble from my ma . . .'

'You're already in trouble – from me. Why are you painting insults on the rectory door?'

As he shook the boy again he was aware that a dog was

77

barking furiously inside the house.

'It wasn't my idea. I was told to do it.'

'Told – or paid? Who said you were to do it?'

The silence that followed was ended by another rough shaking. 'It was Billy . . . Billy Stanbeare.'

'I guessed as much.' The boy had confirmed Sailor's suspicions. 'In that case, you and me will pay Billy Stanbeare a visit.'

Sailor lowered Tim, allowing his feet to touch the ground, but maintaining a firm grip on the boy's collar.

'Billy's not at home,' replied the frightened boy. 'He's at the wrestling championships, over to Launceston.'

'That's very disappointing,' Sailor declared. 'Never mind, we'll still go to his house, so I'll know where to find him when I come looking. Pick up that bucket and we'll be on our way.'

A course of action was beginning to take shape in Sailor's mind, but he would first need to know where Billy Stanbeare lived.

Sailor led the young village boy away from the rectory, his own reason for being there temporarily forgotten. When Alice peered out from behind the curtains at one of the downstairs windows, alerted by Digger's persistent barking, they had already passed out of view and she could see nothing.

Billy Stanbeare was not in the best of moods when he returned from Launceston the next day. He had lost his title to a fisherman from the south Cornwall village of Mevagissey and, with it, a sizeable purse. Accompanying him were three friends who had tried to cheer him on the long walk back to Treleggan, but they had lost too much in wagers for their sympathy to sound genuine.

They were talking of arranging a rematch with the new champion as they cleared an area of fern and bushes and came within view of Billy's valley home – or where the cottage had been when he left for Launceston.

There was nothing on the spot now but a heap of cob and thatch.

The four men came to a sudden halt, disbelief rendering them temporarily speechless. Billy was the first to find his voice, but his spluttering made little sense.

'What . . . ? What's happened?' asked one of the Treleggan men uncertainly.

'That's a bloody stupid question!' Billy raged. 'My home has been knocked to the ground, that's what's happened – and there's no need to guess who's done it. You were all at Widow Hodge's kiddleywink when Alice's navvy threatened to do this.'

'But why?' persisted his friend. 'He said he'd only do it if any of us upset Alice. Nobody's done anything.'

Billy had not told the others of his instructions to Tim Yates. He did not enlighten them now. 'You don't think he was going to wait until he had a reason, do you? Well, he's gone too far this time. You're all witnesses to the threat he made against us – to me, especially. Come on, we'll borrow a pony and cart from my pa and go and find the magistrate.'

'Do we know the navvy's name?' one of the Treleggan men queried.

'His name is Davey, Gideon Davey.' Billy had learned Gideon's name from one of Widow Hodge's serving-girls and it was indelibly imprinted upon his mind. 'It's a name that will soon be on a magistrate's warrant. Mister Davey is going to pay for this. And when he's locked up in gaol we'll run Alice Rowe out of the village too.'

II

Henry Stanbeare refused to believe his son's cottage could have been demolished by navvies until he had seen it for himself. However, once he had visited the scene he was more outraged than his son, unaware of the incident which had prompted such Draconian retribution.

'I'll come with you to see Magistrate Grose,' he said to his son. 'There's no need to bring the others along. He and I have met many times on various committees, he'll not doubt my word. Mind you, I wouldn't have believed such a thing could happen had I not seen it with my own eyes!'

The home of Magistrate Grose was a substantial manor some five miles to the east of the market town of Liskeard and it was dusk by the time father and son reached the house. Magistrate Grose was settling down in front of a cosy fire, preparing to enjoy his first whisky of the evening. Henry and Billy were invited into his study where Grose listened with increasing disbelief to their story.

When they had done, he said, 'Are you telling me that this . . . this labourer, or navvy, whatever you care to call him, actually demolished your cottage during the thirty-six hours or so that you were away? I find that very difficult to believe!'

'He wouldn't have done it by himself,' Billy declared testily. 'He has about fifty navvies working for him. They're a rough lot – most of them criminals, I dare say. They're ready and willing to do whatever Davey tells them.'

Still dubious about Billy's story, the magistrate asked, 'Are you quite certain your cottage was *knocked* down? I have a couple of cob and thatch cottages on the estate. If the thatch is not properly maintained, rain can get into the walls and they are likely to collapse without warning.'

'So they might,' Billy retorted heatedly, 'but when they do they don't usually take granite chimneys with them.'

Aware that Billy was losing his temper, Henry said hastily, 'The cottage has been totally demolished, Mister Grose. I've seen it with my own eyes. It's the work of men, not of nature.'

'Very well,' the magistrate said patiently, 'let us assume for the moment that such astonishing vandalism is the result of a criminal act. Do you have any proof that it was perpetrated by this Gideon Davey?'

'I can give you the names of more than twenty witnesses who heard him threaten to do exactly what's been done,' Billy said, still angry. 'There's no doubt about it at all.'

'Very well, I will issue a warrant and have Davey arrested. You will hear from me in due course.'

There was a note of resignation in the magistrate's voice. He was aware that a number of warrants were outstanding against navvies, but his constables had been unable to serve them. In most cases they had received no co-operation from the sub-contractors building the railway. Even when co-operation was forthcoming his constables were reluctant to take action. Any attempt to arrest a navvy meant putting their own lives in very real danger. However, this was a serious matter and although Henry Stanbeare did not possess much actual influence in the county he was acquainted with a number of men who did.

The loss of his home was a greater blow to Billy's pride than to his pocket. The cottage had been little more than a hovel and scantily furnished. It was also so damp that Billy kept any worthwhile belongings, including his best clothes, in the room at his father's farmhouse set aside as 'Billy's room'.

Although his home had been destroyed, Billy thought it would be a price worth paying in order to have Gideon put away in prison. He believed it would restore his position in the community as the natural leader of the young men of the area. It would also leave him free to take his revenge on Alice for being the cause of his humiliation in the first place. When the time came he would do more than have names painted on the rectory door.

Many years before, when Alice was a young girl new to her work at the rectory, he had given her an opportunity to become an accepted, if not respected, member of the Treleggan community. Aware of what she would need to give to Billy in return for his support, she had spurned his offer – and paid the price.

Now she was known to have entertained a navvy, he would take her. If she dared to lodge a complaint against him he would say she had encouraged him – and no one would disbelieve him.

Having Alice was a prospect that Billy looked forward to with considerable pleasure.

* * *

For two days Billy and his father waited to hear from the magistrate that Gideon had been arrested. On the Tuesday, Billy visited his demolished cottage in the moorland valley and found Magistrate Grose and a constable inspecting the ruins.

Hurrying towards them, Billy asked eagerly, 'Have you come to see for yourself what Davey's done? Does it mean you've arrested him?'

'No – to both questions,' Magistrate Grose replied. 'I am here to see for myself whether or not a crime has been committed. To be honest with you, Mister Stanbeare, I am not convinced.'

Billy looked at the magistrate in disbelief. 'What do you mean, you're not convinced? What more do you need? You've got the proof right in front of your eyes. This heap of rubble was my home when I left it last Saturday. When I returned on Sunday this is what I found.'

'Ah, that's what happens with these cob-built houses when they're not kept up together,' said the constable sagely. 'I've seen more than one collapse when we've had heavy rain. Looking at this one, you can see where the mud's been washed down towards the stream.'

'We've only had that sort of rain in the last couple of days,' Billy retorted contemptuously. 'Since my house was demolished by Davey. I've already told you I've got more than twenty witnesses who heard him say what he'd do. I'd have thought you'd have arrested him by now.'

'You could produce a thousand witnesses but it would make no difference,' Magistrate Grose said unsympathetically. 'Gideon Davey was most certainly not involved in what has happened here. He left the country a week before you reported the destruction of your cottage.'

'Who told you that?' Billy demanded angrily. 'Some of his navvies, I suppose. You're a fool to listen to them. They're all as bad as he is.'

The magistrate was not used to being spoken to in such a manner and he showed his displeasure. 'It seems you are in the habit of questioning everyone's judgement except your own, Mister Stanbeare. I had no need to question any navvies. I received my information from the engineer in charge of the whole railway building programme. Mister Davey has accompanied a very well-known and respected contractor to a foreign country to help estimate the cost of a new railway – and from what I have heard of him, he is hardly the type of man to knock down other people's homes for the fun of it. I repeat, it is my considered opinion that your cottage was probably in a poor state of repair. As a result, rain found its way into the main structure and caused it to collapse. There is no evidence of any criminal act involved, so I will bid you good day, Mister Stanbeare.'

Turning his back on Billy, he said to the constable, 'Come along, I have other matters to attend to. We have both wasted quite enough time here.'

That evening Billy returned to his father's farmhouse still smarting at the manner with which the magistrate had dismissed his assertion that Gideon had been responsible for demolishing his cottage.

It also gave Billy much food for thought. It meant that Gideon had kept another of his promises. Even if he was not in the area himself, any interference with Alice would result in drastic retaliation by those navvies who were.

Young Tim Yates had been adamant that no one had seen him in the rectory grounds, but he had lied to Billy about his second abortive attempt to daub his message on the rectory door. He said he had gone there, but realised there was someone standing guard in the garden and had fled, succeeding in losing the man who chased him.

Billy accepted the boy's story, believing that whoever Gideon had detailed to protect Alice had automatically assumed he was to blame and carried out the orders of the ganger and demolished his cottage.

He told his father of his beliefs that evening, leaving out the part that Tim Yates had played, blaming Alice for encouraging Gideon in the first place.

'Just you bide your time, Billy,' said his father smugly. 'I don't think we will have to put up with Alice for very much longer. I had a letter from the rural dean this morning. A new rector has been appointed. He'll be arriving in Treleggan tomorrow.'

'How will that affect Alice?' Billy asked. 'A new rector will need servants and she and Ivy are already there.'

'She won't be there for long,' Henry declared. 'The man who's been appointed has just returned with his wife from a parish in India. They're bringing two Indian girls with them as servants. I doubt if a girl with the morals of Alice Rowe will be acceptable in such a household. The new rector is coming by rail to Plymouth, and the rural dean has asked me to hire a coach and meet him and his wife and servants there. He said I should tell Alice of their expected arrival, but I've decided to keep the news to myself. Who knows, when the rector arrives he might find some of the navvies in the rectory with Alice. That should ensure her dismissal on the spot!'

Billy agreed it was most probable that the arrival of a new rector who was bringing in his own servants should result in the dismissal of Alice, but he did not share his father's elation. Gideon was no longer in the country but he had already shown he was capable of keeping the promise he had made when he humiliated Billy and the other young men of Treleggan.

It was not beyond the realms of possibility that the Stanbeare farmhouse would suffer the same fate as Billy's cottage if it was suspected that Henry was the cause of Alice's dismissal from the rectory.

9

I

The week had begun badly for Alice, but it was to become dramatically worse.

She was deeply upset by Gideon's failure to visit her on Sunday. She had thought about him a great deal since enjoying the surprise birthday treat he had arranged for her, and was aware he had been very disappointed by her reaction to his totally unexpected proposal of marriage. She also realised that her response would have been different had there been time to think about it; had she not been so upset by what they had found at the rectory on their return.

She thought, bitterly, that if it had been the intention of the perpetrator to hurt her, he had succeeded beyond all his possible expectations.

Had Gideon come calling, as he had promised, she would have told him that once the events of that night were over she had been able to consider what he had said. Indeed, she had thought of little else all the week whether during her working hours or in the quiet of her room.

Alice still believed they would need to get to know each other better in order to be absolutely certain of their hearts, but there was no doubt in her mind that she reciprocated his feelings for her. She intended telling him so.

As Sunday approached she became increasingly excited at the thought of seeing Gideon again, although something happened after darkness fell on Saturday evening to provide a temporary distraction.

Digger had begun it by making a fuss in the hallway, in the way he always did when someone came to the rectory door. Suddenly excited at the thought that Gideon might have decided to pay her a call earlier than arranged, she hurried from the kitchen.

It was dark outside and she hesitated a moment before opening the front door. She was about to call out and ask whether it was Gideon when she heard voices – and one of them sounded angry. Neither voice belonged to Gideon and she changed her mind about opening the door. Instead, she quietly slipped the bolts home so the door could not be opened from the outside. Then she hurried to the late parson's study, which overlooked the front lawn, and peered out of the window, but she could neither see nor hear anything.

Returning along the passageway to the hall, she paused once more to listen. It was quiet outside now and she decided it must have been sounds from the village she had heard. Voices tended to carry when a wind was blowing from that direction. Nevertheless, she left the front door securely bolted.

It was not until early on Sunday morning when she let Digger out of the house that she discovered there *had* been someone outside. She found a paint brush lying on the pathway at the front of the rectory.

The find sent a chill through her. Her thoughts of Gideon had pushed events of the previous Saturday evening to the back of her mind but now they flooded back. She believed the unknown dauber had returned, this time with a companion, but had been frightened off by the barking of Digger. It seemed that yet again whoever it was had succeeded in turning her thoughts from Gideon.

She awaited his visit with even more urgency now, convinced he would be able to put this latest incident into perspective for her.

As time passed and he did not arrive, her mood changed to one of concern, and then despondency. Reassessing the events of his last visit, she began to think that Gideon had decided not to come because of the lack of enthusiasm she had displayed when he told her how he felt about her.

She tried to tell herself that Gideon was not a man who would give up so easily if he really cared about her. Nevertheless, when darkness fell and he had still not arrived, she could fool herself no longer.

Bitterly, she told herself it was all her own fault. She had finally met a man who meant something to her and had managed to make him feel rejected.

His failure to continue with their relationship had other, albeit less important implications. When the villagers became aware that Gideon's visits to her had ceased they would either believe he had succeeded in taking what he wanted from her and was no longer interested, or think he had bowed to the opposition of Billy Stanbeare and his friends. Either way, they were certain to make life even more difficult for her.

Alice could not know that Sailor had been so jubilant at catching Tim Yates and making good Gideon's threat to demolish the house of anyone picking on her that it was not until Sunday night, after a day of celebration, that he remembered the message he should have delivered.

By then it was far too late to consider walking to Treleggan and calling on her. Nor was there likely to be an opportunity during the coming week. He would be kept busy organising the gang for the new task they were contracted to perform, farther along the track.

His mission would need to wait until the following weekend. He would apologise to Alice and tell her the reason why Gideon's message had not been delivered on time. He felt certain she would understand.

* * *

On Wednesday evening, Alice was tidying the kitchen prior to locking up and retiring to her room. She had borrowed a book from the late parson's library, intending to read a chapter or two before she went to bed. It was raining heavily outside and if she remained downstairs it would be necessary to light lamps about the house.

She was alone in the rectory, Ivy having put in an appearance for only an hour earlier that morning. Saying there was nothing for her to do and that it looked like rain, she had returned to the house she shared with her sister.

Suddenly, Digger began barking and refused to stop when she told him to shut up. Believing one of the village dogs had found its way into the rectory garden, Alice decided to go to the front door and shoo it away if it was close to the house.

She was only halfway along the passageway that led to the hall when the front door was flung open and a woman stumbled inside the hallway, closely followed by two dark-skinned girls, a slightly bemused man wearing clerical gear – and Henry Stanbeare.

II

The Reverend Emmanuel Bushell and his wife, accompanied by two Indian maidservants, stepped from the train at Plymouth railway station and were given an effusive welcome by Henry Stanbeare, who introduced himself as Treleggan's senior churchwarden.

When porters had gathered the considerable amount of luggage which accompanied them, Stanbeare led them to the waiting carriage, saying, 'I am delighted to be the first to welcome the new rector of Treleggan and his lady. I am sure you will be very happy in the parish and hope you will remain with us for at least as many years as our last, dearly loved incumbent.'

'I hope Treleggan is warmer than Plymouth,' Harriet Bushell

commented, before her husband could reply. 'Reverend Bushell accepted the living because it was our understanding that Cornwall is warmer than anywhere else in the British Isles. After spending so many years in India I feel the cold dreadfully.'

Henry Stanbeare did not think this was the moment to tell her that Treleggan was a thousand feet above sea level and surrounded by some of the bleakest terrain to be found anywhere on Bodmin Moor. Instead, he said glibly, 'I have no doubt the rectory servants will have the house nice and warm, ready for your arrival – although I am afraid you will find the present housemaid, Alice, somewhat unreliable. It's the result of not having a master in the rectory since the death of poor Parson Markham.'

'She'll soon learn I do not tolerate laziness,' Harriet said curtly. 'As an English girl she will be expected to set an example to my Indian maids.'

Henry had quickly realised that Harriet was the dominant partner in the Bushell family and it was to her that he fed a number of false 'facts' about Alice on the long journey to Treleggan. He told her of Alice's background, adding a great many embellishments of his own, declaring that Parson Markham had employed her against his advice.

'The parson was a warm-hearted and generous man,' Henry said hypocritically. 'I fear that, sadly, there are those in this world who take advantage of such kindness.'

'She will not take advantage of Emmanuel,' the rector's wife said firmly. 'He is a true Christian and far too generous and forgiving, but he has me to ensure such generosity of spirit is not abused . . . but I fear it is becoming colder. Will you please call out to the driver and ask whether he carries any rugs on the coach. I am thoroughly chilled.'

It was a complaint the rector's wife had voiced many times since leaving the train. She felt so cold she had refused to leave the carriage during the passage of the ferry across the River Tamar, saying the wind was far too keen.

When they first set off, the two Indian girls had been ordered to ride on the outside of the coach. However, when it began to rain, Emmanuel Bushell insisted that they be allowed to join the others inside the carriage, Harriet ordering them to remain silent and 'not be a nuisance'.

Henry felt that with only a little priming the new rector's wife was capable of dismissing Alice on the flimsiest of pretexts. While pretending to be sympathetic to the rectory servant, he managed in a subtle way to blacken her character in the eyes of Harriet Bushell.

'. . . Alice, this is Reverend Bushell, the new rector, and Mrs Bushell. They have had a long journey to Treleggan and are hungry and cold. I told them there would be fires in their room to revive them and that by the time they had washed and changed you would have a meal on the table for them.'

Alice looked at Henry in dismay. 'But . . . I knew nothing about a new rector arriving today! There's hardly any food in the house, except for bread and cheese, and no fires lit, except in the kitchen.'

Henry feigned astonishment. 'No food . . . or fires? But you knew . . . Oh dear, this is most unfortunate! Reverend and Mrs Bushell have had a very long journey.'

'Had I known a new parson had been appointed and was arriving today I would have had everything ready, but I did *not* know.' Alice realised that Henry had deliberately neglected to tell her and she was angry.

'You will please remember your place, Alice!' Harriet Bushell snapped. 'There must be *something* more than bread and cheese in the house.'

'There isn't,' Alice said bluntly. 'But Mister Stanbeare has a farm. I'm sure he can produce something. While he's fetching it I'll get fires going in the downstairs rooms and in the bedroom, but first I'll build up the kitchen fire. When he returns it won't take me long to cook something for you.'

At that moment, Digger, who was in the corner of the scullery, just beyond the kitchen, began barking furiously again.

'What's that?' Harriet demanded.

'It's Digger, ma'am,' Alice replied. 'He was Parson Markham's constant companion. He was with him when he died . . .'

'I don't care whose companion he was,' Harriet snapped, 'I will not have a dog in my house. They are filthy creatures. Put it outside, where it belongs.'

Alice was horrified. 'But Digger's always lived in the rectory. He's an old dog now—'

'Did you hear me? Put the animal outside this instant – and it is not to come inside the house again.'

Alice would have continued to plead Digger's cause, but she saw the smug expression on Henry's face and knew he would be delighted by just such a confrontation.

'Yes, ma'am, I'll put him out now.' Opening the scullery door, she caught the small Jack Russell as he rushed into the kitchen. Carrying him outside, she hurried to the wood shed. Hastily arranging a couple of sacks on the floor, she said, 'You stay here, Digger – and be quiet. I'll come and feed you in the morning.' Hoping she had settled him for the night, she gathered a number of logs in her arms and hurried back to the kitchen.

In the hall of the rectory Henry was leaving to fetch food and Harriet was saying to him, 'I do apologise, Mister Stanbeare. It is unforgivable for a servant to virtually demand that you provide us with food because she has failed to carry out her duties. Unfortunately, unless you do I fear we must go hungry.'

'I shall be happy to provide what I can,' Henry said, 'but you can see I was not exaggerating when I told you that Alice is unreliable. Had poor Parson Markham not been so kind-hearted she would never have been kept on here.'

'She will not find *me* so easy-going,' declared a tight-lipped Harriet. 'She will do things my way – and do them well. Once my two Indian girls learn where everything is Alice will realise how things *should* be done. If she fails to meet the same standards, she will go.'

Henry Stanbeare went away from the rectory feeling very pleased with himself. He believed it would not be long before Alice departed from Treleggan.

III

The next few days were unhappy ones for Alice. Nothing she did satisfied Harriet Bushell. Alice believed she might have got along with the Reverend Bushell, but he was dominated by his overbearing wife and would do nothing likely to incur her wrath.

Ivy Deeble, the rectory's ageing housekeeper, fell foul of Harriet within minutes of meeting her, when the new incumbent's wife learned that she was not in the habit of keeping detailed accounts of household expenditure.

That afternoon, after constant criticism of her methods, Ivy told Harriet she was not prepared to put up with any more of her 'nonsense' and walked out.

In truth, she had been considering retirement, even before the death of the Reverend Markham. The arrival of Harriet Bushell had decided matters for her.

During the furore caused by the housekeeper's resignation, Alice was able to take out food for Digger. The small dog was now ensconced in the garden shed. It was far enough from the house for his barking not to give his hiding-place away, but Alice knew she could not keep him there for ever.

She now felt increasingly alone and isolated in the rectory. Harriet found fault with almost everything she did and the two Indian servants conversed in their own language, excluding her from their conversations. It seemed to Alice they were aware their employer disapproved of her and were frightened of incurring her displeasure by appearing to be kindly disposed towards their fellow servant.

As a result, there was no one with whom Alice could discuss her problems. Matters came to a head on the Saturday, only a few days after the Bushells' arrival.

Emmanuel Bushell was to give his first sermon the following day in Treleggan church, to a congregation made up of largely indifferent but extremely inquisitive parishioners. They would attend church to see whether their new parson's power of oratory matched the 'hellfire and brimstone' sermons for which the local Methodist minister was renowned. They would also be hoping to catch a glimpse of the two Indian servants, whose colourful garb was the wonder of those parishioners who had already seen them.

Aware of the importance of the first service he would be taking in his new parish, Parson Bushell had been in the study for perhaps twenty minutes before he suddenly stomped angrily into the kitchen.

Harriet was supervising an Indian-style meal being prepared by one of the servants, while Alice scoured the pots and pans that had seen use during Parson Markham's tenure, and Harriet had said looked as though they had not been properly cleaned for years.

Scowling, Emmanuel Bushell complained to his wife, 'There is a dog barking incessantly somewhere near at hand. It is destroying my concentration. Do you think someone might do something about it?'

Realising it must be Digger, who was becoming increasingly bored by having to spend most of the day and night shut in the garden shed, Alice said quickly, 'I'll go and find out where it is and make sure it's kept quiet.'

She had almost reached the kitchen door when Harriet called sternly, 'Alice.'

Filled with foreboding, Alice halted, her hand on the door handle. 'Yes, ma'am.' She thought she knew what was to come – and she was right.

'What did you do with that animal I told you to put out of the house when we first arrived at the rectory?'

'I put it out, ma'am, just as you told me to.'

'*Where* did you put it?'

Alice turned to face Harriet and the two women locked glances. 'Digger was Parson Markham's dog, ma'am. He was very fond of it. Digger proved his loyalty by staying out on the moor all night with him when he died.'

'Answer my question, girl,' Harriet snapped angrily. 'Where is the dog now?'

Alice realised it was no use lying to the other woman. She

had only to go outside to realise where Digger was being kept. 'He's shut in the garden shed, ma'am.'

'So you deliberately disobeyed me!' Harriet was furious.

'I put it out of the house, ma'am, as you told me to do.'

'Don't prevaricate. You will go and put it out of the garden – this instant!'

'Please, ma'am, can't I take it somewhere . . . somewhere where it won't disturb the parson? It's a very old little dog.'

'I do feel that perhaps we owe that to Reverend Markham's memory, dear,' Emmanuel Bushell said to his wife, aware he had unwittingly prompted a confrontation between the two women. 'During my service tomorrow I will be praising his work for the parish. It would be hypocrisy were I responsible for harm coming to something that meant so much to him.'

'I will not have a servant disobeying my orders, Emmanuel,' Harriet said. 'However, in deference to your wishes . . .' Glaring ferociously at Alice, she said, 'Very well, you may take the dog off somewhere – but you will make up for the time you are wasting by working late this evening. And should the dog ever return to the rectory again I will have the gardener shoot it. Reverend Bushell is now the rector of Treleggan and there will be no smelly, dirty animal in the house while *I* am here, is that understood?'

'Yes, ma'am . . . thank you, ma'am.' Much relieved, Alice ran out of the rectory without waiting to put on a coat, even though it was cold and damp outside. She was afraid Harriet Bushell might change her mind once the rector left the kitchen.

Digger was delighted to see Alice and when she released him from the shed he danced about her, jumping up and barking excitedly. Picking him up, she hugged him and he wriggled in her arms, trying to lick her face.

'Poor Digger, your life has been turned upside-down too, but you don't understand, do you? Never mind, we'll see if Ivy will look after you for a day or two, while I try to think what we can do with you.'

The ex-housekeeper was reluctant to take on the respon- sibility of caring for the small dog. Her sister too expressed doubts, her concern being for her cat, which was even older than Digger. However, Harriet's threat to have the Jack Russell shot proved the deciding factor. It was felt that with a little planning, cat and dog could be kept apart – but the elderly sisters impressed upon Alice that it must be a purely temporary measure, for a few days only.

With Digger's immediate future settled, Ivy sighed, 'Poor Parson Markham. He'd be turning in his grave if he knew how things were at the rectory. Are they likely to get any better for *you*, Alice?'

Alice shook her head. 'Not while she treats me like a scullery-maid. The two servants she brought with her from India are doing my work. I believe she's hoping I'll have enough of it and leave.'

'Well, if you leave in a hurry don't forget to take the parson's dog with you.' The reminder came from Ivy's sister, who appeared to be having second thoughts about giving Digger a home, albeit on a temporary basis. 'I won't see a healthy and faithful animal put down without good cause, but there's no place for a dog here. You must make some other arrangement as soon as you can.'

In fact, Digger would remain with Ivy and her sister for less than twenty-four hours, by which time Alice would no longer be employed at the Treleggan rectory.

IV

Upon her return to the rectory after finding Digger a temporary home, Alice discovered that Harriet had unearthed more cooking pots for her to scour and she was kept busy for the next few hours carrying out the dirtiest jobs that could be found.

At the end of the day, when the Indian servants had completed their duties, Alice was ordered to scrub the flagstone floor of the kitchen. She set about the task without comment, even though it had been scrubbed only that morning.

Only half the task had been completed when there came a knock on the door which led from the kitchen to the garden.

Alice's immediate thought was that it might be Gideon, in spite of the time that had elapsed since she had last seen him. Her heart beating faster, she hurried to the door. She was about to open it when she remembered another, less welcome caller who had painted 'whore' on the front door.

'Who is it?' she called softly.

'You don't know me, miss. I'm Sailor Smith, one of Gideon's men. I have a message from him.'

Alice opened the door hurriedly. It was raining outside and the very large man standing there was soaked through.

'You'd better come just inside, but keep your voice down, there's a new rector and his wife in the house. What is Gideon's message? Is something wrong?'

Alice fervently hoped that Sailor had come to say that Gideon would be calling on her, but his next words dashed her hopes. 'He's had to go away, miss. Gone abroad to the Crimea, with the contractor. He's likely to be gone for a couple of months or so. I was supposed to deliver his message a week ago, but . . . well, I wasn't able to.' He thought it best not to mention the events of that particular night. 'Gideon was very unhappy about not being able to come and tell you himself, but he said I was to say he'd write as soon as he could . . .'

At that moment the door linking the kitchen with the rest of the house opened and Harriet Bushell entered the kitchen.

At sight of the bedraggled man towering over Alice, she let out a shriek. 'Wh . . . who are you? What are you doing here, in my home?'

'He brought a message for me, ma'am. It was raining so much outside I invited him to step into the kitchen. He's only just arrived . . .'

'*You* invited him in? How dare you!' Recovering from her initial shock, Harriet looked Sailor up and down with undisguised distaste. 'Who is he? Is he a local man?'

'No, ma'am.' Sailor spoke to her for the first time. 'I work on the railway. As Alice said, I am merely delivering a message from my ganger.'

'You're a *navvy*! I was warned that Alice was in the habit of entertaining navvies in the rectory when there was no one here to put a stop to such goings-on. Well, there's a new rector here now and such visits are at an end. You will leave the rectory – *immediately* – before I call my husband.'

'Of course, ma'am.' Sailor spoke with careful politeness. 'I hope I haven't made any trouble for Alice. I was only delivering a message – and I doubt very much whether any other navvies have called here.'

'I am not interested in your opinion,' Harriet retorted angrily. 'I told you to leave my house.'

'Of course, ma'am.' Opening the door, Sailor stepped out into the rain and the darkness, but before he disappeared, he said softly to Alice, 'I'm sorry if I've got you into trouble, but remember, Gideon said he would write.'

Closing the door behind the huge navvy, Alice turned and braced herself for Harriet Bushell's wrath. It was turned upon her immediately.

'I have put up with more than enough from you, young lady. This is the last straw. Mister Stanbeare told me what I might expect from you, but I gave you the benefit of the doubt. I was wrong.'

'Henry Stanbeare has never had a good word to say about me,' Alice said bitterly. 'He was saying things about me before he'd even met me.'

'Don't answer me back!' Harriet snapped. 'It is no use trying to put the blame on Mister Stanbeare for your shortcomings. I have experienced them for myself. I have had enough of you. You will pack your things and leave the rectory tomorrow.'

Alice had known her employment at the rectory was in serious jeopardy, but she had not expected the end to come as swiftly as this. The blood drained from her face and she felt suddenly sick. 'But . . . where will I go?' she said in a stricken voice. 'I know no one.'

'That is not my concern,' Harriet said callously. 'I have no doubt you will find somewhere more suited to your behaviour, probably among the navvies you seem so fond of. Now . . .'

Alice did not wait to hear more. She could feel tears burning her eyes and she had no wish to give Harriet Bushell the satisfaction of seeing her cry.

Behind her in the kitchen, Harriet looked in distaste at the bucket, scrubbing brush and half-scrubbed floor, then went in search of an off-duty Indian maid.

Alice did not put in an appearance the next morning. As this would have been her day off anyway, it excited no comment – although Emmanuel Bushell complained of the quality of the breakfast cooked and served by the Indian servants.

'You will need to get used to it,' Harriet replied irritably. 'We can't afford to employ a cook – even if we could persuade one to come to such an out-of-the-way place as this.'

'Alice is a very acceptable cook.'

Harriet took her husband's words as an accusation and she said sharply, 'Alice is also an unreliable servant, with moral standards that have no place in a rectory.'

'We really only have Henry Stanbeare's evaluation of her, dear, and Alice told you he did not like her. I visited Ivy Deeble on Friday. She said he has always had a down on the girl. It seems her mother once made him look very foolish and he has taken his spite out on Alice. She was also adamant that she and Alice were never told of our imminent arrival . . . at least, that's what I was told,' he added hastily when he saw his wife's expression harden.

'Ivy Deeble is as deaf as a post,' Harriet declared, 'She would not have heard if he *did* tell her, as I have no doubt he did, and his opinion of Alice was confirmed by the presence of that navvy in the kitchen last night.'

'She gave you an explanation for that, dear – albeit one you felt unable to accept. Perhaps we should have found out a little more from her. That would have been the charitable thing to do. After that initial misunderstanding I feel she performed her duties quite adequately.'

98

In spite of all she had said to Alice, Harriet was forced to admit that what her husband had said was quite true. Henry Stanbeare's stories about Alice had turned her against the maid before the two women had met, but Alice had proved herself to be a hard worker.

Sensing his wife's indecision, Emmanuel Bushell said, 'On a more practical level, it would be difficult to do without the girl, at least until we know where everything is in the rectory, and where all the necessities of life may be purchased.'

This latter consideration was something that had occurred to Harriet only that morning when she had needed to supervise the lighting of the kitchen fire and the preparation of breakfast. 'Very well, Emmanuel, I will be lenient, as you suggest, and give the girl another chance. But I will say nothing to her until after this morning's service. She can stew until then. Alice will not leave without coming to me and asking for whatever wages she may believe she is due.'

V

After the altercation with Harriet Bushell, Alice shut herself in her room and cried bitterly for more than an hour. Then, taking a firm hold on herself, she tried to think of what she would do, and where she could go.

Contrary to Harriet's assumption, Alice had no need to approach her for money. Rural Dean Brimble had paid Alice in advance, giving her sufficient funds to feed herself and Ivy and keep the rectory heated for at least another month.

She would return what was left of the money to the Reverend Brimble, who lived in Bodmin, and appeal to him for help in finding new employment. But first she would collect Digger from Ivy. Alice thought she might also pay a call on Widow Hodge on her way to Bodmin, to ask if she had any suggestions about possible employment.

Sadly, the disreputable old woman was the only person in

the vicinity of Treleggan with whom Alice felt she could talk about what had happened.

Alice departed from the rectory when she knew the service had begun in Treleggan church. Most of the villagers would be attending the service in order to get a good look at the Reverend Emmanuel Bushell. After the service they would discuss the power of his oratory and express a firm opinion on whether or not he was acceptable to the insular moorland community.

Carrying a bundle that contained her pitifully few personal possessions, Alice walked through the churchyard on her way from the rectory. As she passed by the centuries-old building the hymn *O come, O come, Emmanuel* was being sung and it brought tears to her eyes. It was obviously being used as a reference to the Reverend Bushell's Christian name, but it had also been Parson Markham's favourite hymn.

Skirting the quiet village, Alice made her way to the cottage of Ivy and her sister, both of whom had decided to boycott the church service. After only a brief explanation, she left with their sympathy and good wishes and with Digger trotting happily at her side, on the end of a string lead.

It was still early when she arrived at the moorland kiddley-wink and there were no customers to disturb her talk with Tabitha Hodge. They would not begin to arrive until midday. When Alice told her she had been dismissed from the rectory the old widow invited her inside and listened in silence to Alice's story.

When it had been told, Tabitha Hodge said, 'It sounds to me as though Treleggan has finally got the parson the folk there deserve – though if his wife rules the roost he's not likely to remain there for long. She'll miss the social life they'll have enjoyed in India and will upset the women in the village as soon as she opens her mouth to them. But why have you come to tell me about it?'

Alice was not entirely certain herself. 'I don't know, except . . . there was no one else I felt I could talk to. I thought you might have some work I could do. Cleaning, or cooking . . . or something.'

'I might be old, m'dear, but I can still do my own cleaning and cooking. As for the "or something" . . . I have two girls working here who do all that's asked of them, by whoever wants them. They wouldn't welcome competition. Besides, it isn't the sort of thing you'd want to do, whatever might be said about you in Treleggan. What about that young man who's sweet on you – the navvy? Have you told him what's happened?'

'I can't,' Alice said unhappily. She had thought a great deal about Gideon while she was tramping across the moor to the kiddleywink and realised how different things might have been had he not gone away. 'The navvy who came to the rectory last night was there to tell me that Gideon has had to go abroad with the railway contractor. He won't be back for a long time. He's promised to write, but unless I stay somewhere near I won't ever get his letter. I . . . I just don't know what to do.'

'You said just now you were on your way to see the rural dean, at Bodmin. Perhaps he'll be able to help. He certainly wouldn't approve of you working here. I don't think he approved of Parson Markham coming to see me as often as he did.' Aware how unhappy and desperately alone Alice was feeling, Tabitha asked, 'Have you eaten this morning?'

Alice shook her head. 'I had nothing last night, either. I was scrubbing the kitchen at the time when I'm usually getting myself something to eat and didn't feel like eating after Mrs Bushell dismissed me.'

'Come into the kitchen and I'll cook you some bacon and a couple of eggs. It's a long walk to Bodmin, you'll need something inside you.'

In the kitchen, Alice helped Widow Hodge, at the same time marvelling at the old woman's energy and ability. When the meal was ready Alice sat down to eat and Tabitha sat facing her across the table. Suddenly and unexpectedly, she said, 'You are very like your mother, girl.'

A forkful of food halted just short of Alice's mouth and, astonished, she said, 'You knew my ma?'

Tabitha nodded her head. 'I knew her before you were born – and when she was carrying you too. She used to come up past here to meet your father. Sometimes, when I was out gathering wood up in the valley I'd see them walking hand in hand. They made a handsome couple. It was sad the way things turned out for her. Sad for him too, of course. He was a nice boy.'

'You knew my father too!' The fork was lowered to her plate, the food on it untouched. 'What was he like? Ma never liked talking about him.'

'I can see some of him in your looks,' Tabitha said. 'Especially about the eyes. I always said he was far too much of a dreamer to be working in a mine. He'd have married your ma, you know, had it not been for the accident. She certainly didn't deserve to be treated the way she was by them at Treleggan. She did no more than most girls do when they find the man they want to marry. It was Henry Stanbeare who set everyone against her. You know your ma had been brought up by an aunt?'

Alice nodded. She was aware that her grandparents had both been carried off in an epidemic that took the lives of a great many Cornish men and women.

'Yes, your aunt was so upset by what Henry Stanbeare stirred up against the family that she left Treleggan without telling anyone where she was going, while your ma was in prison with you. Henry Stanbeare has a lot to answer for. He's a vindictive man, Alice, and never forgives a slight, whether it's real or imagined.'

'What about my pa's folks?' Alice asked eagerly. She had often toyed with the idea of finding them, to see if there was a chance they might acknowledge her as their granddaughter.

Tabitha shook her head. 'Your pa was one of three mining brothers. The other two went off to America and made good, or so I heard. Whether they did or not, they sent for their ma and pa to join 'em. This would be some twelve or fifteen years since. They'll all be doing well out there now, I've no doubt.'

'So I really *am* all alone,' Alice said, with resignation rather than bitterness, 'but there's nothing new in that.'

'No, you're not alone,' Tabitha said. 'You have that young

man who gave Billy such a beating – and whatever was said about it, he did it on your account. He didn't give me the impression of a man who would build up a girl's hopes only to let her down.'

'I don't think he would,' Alice agreed, 'but he's not in the country, or I'd go and find him. By the time he comes back I could be anywhere.'

'He'll come looking for you, you mark my words,' Tabitha said firmly. 'When he does I'll make certain he learns that you went to Parson Brimble for help. If you're in work, no matter where it is, your young man will find you. In the meantime, don't give up hope that things will eventually turn out right for you. I've a feeling in my bones that they will.'

VI

During the five-mile walk to Bodmin, Alice tried to take heart from Tabitha Hodge's words. It was not easy. Rain began to fall before she had covered half the distance and as her bundle of belongings grew wet it got heavier. In addition, her poke bonnet became saturated, causing the brim to sag wearily over her forehead.

Even the stoic Digger found the weather hard to accept. After he had staged a stubborn sit-down on two occasions, Alice picked him up and cuddled him in the crook of her free arm.

In this wet and bedraggled state she reached Bodmin. After enquiring from a townsman of its whereabouts, she arrived at the large house occupied by the Reverend Harold Brimble.

Somewhat apprehensively she tugged at the bell-pull. After what seemed an age, it was opened by a uniformed maidservant. The girl looked with incredulity, bewilderment and a degree of sympathy at the bedraggled girl standing on the doorstep, clutching a small dog and a bundle of wet possessions.

'I've called to speak to Reverend Brimble,' Alice said, conscious of the maid's uncertainty.

Finding her voice, the housemaid asked, 'Who shall I say is calling?' adding, hesitantly, 'Is he expecting you?'

'No, but tell him it's Alice, from Treleggan rectory.'

Still uncertain of this rain-sodden caller, the maid said, 'Wait here, and I'll see if the reverend will see you.'

With that she pulled the door to, leaving a narrow gap through which Alice could see into the dry and carpeted hall.

After a short wait, she heard voices, but when the door was opened it was a homely-looking woman who stood in the doorway, not the rural dean. Staring at Alice in dismay, the woman said, 'My dear soul, you are soaking wet! And that poor little dog too. Come in, my dear. Come to the kitchen where it's warm and have something hot to drink. I am Mrs Brimble. My husband is working in his study.'

Addressing the maid who had opened the door to Alice, she said, 'Tell the master that Alice is here . . .' and then, returning her attention to Alice, she went on, 'We must get you somewhere warm before you catch your death of cold – if you haven't already picked up a chill!'

Alice had gone to the home of the rural dean because she could think of no one else to whom she could appeal for help. Yet as she approached Bodmin she had begun to entertain doubts about the reception she would receive at the home of the senior cleric. After all, she was only a servant – and one who had been dismissed by a clergyman's wife. It in no way made her the responsibility of the Reverend Harold Brimble.

By the time she reached Bodmin she had almost convinced herself she would be turned away and left to fend for herself in a town where she knew no one. The unexpected kindness of Beatrice Brimble's welcome was almost too much. She experienced such a feeling of relief she could have broken down and wept. Not until she reached the kitchen was she aware she was shaking – and it was not entirely because she was cold.

Aware of Alice's distress and uncertainty, Beatrice Brimble pretended not to notice. Sitting Alice down close to the kitchen fire, she said, 'I'll put the kettle on and let Mary make you a cup of tea when she returns. I was going to suggest you change into a dry dress, but by the look of your bundle I doubt there's much dry in there. Let me take the dog from you. It's poor Parson Markham's dog, isn't it? I seem to remember his name is Digger. Arnold Markham once brought him to a diocesan meeting, a year or so ago. Despite his protests, he was forced to leave Digger here for the duration of the meeting. We got on very well, didn't we, Digger?'

The small dog's response was to growl as she reached out for him. Refusing to be intimidated, Beatrice said, 'Now, now! We'll have none of that. Besides, you know you don't mean a word of it.'

She took the wet dog from Alice and, after a token protest, Digger's stump of a tail twitched. When the rural dean's wife placed him close to the fire, he shook himself vigorously before putting his nose to the floor and setting off on a journey of exploration round the room.

At that moment the Reverend Harold Brimble entered the kitchen. A tall, distinguished man, he showed immediate concern for the rain-drenched girl. 'My dear, you are soaked through! Whatever are you doing coming to Bodmin on such a day?'

Looking up at him, Alice once more felt ridiculously close to tears. 'I've been dismissed from the rectory,' she explained. 'Mrs Bushell dismissed me. Ivy's gone too. She left on Parson Bushell's first day.'

Harold and Beatrice Brimble exchanged glances. The appointment of the Reverend Emmanuel Bushell to the living of Treleggan had not been made through the usual channels. It had come about as the result of a personal request from the Bishop of Exeter. Bushell was the relative of a friend. Returning from India because of serious troubles in his parish there, he was in urgent need of a living. The bishop thought Treleggan and the clergyman from India might suit each other.

Harold Brimble had expressed doubts to his wife about such an appointment. Treleggan was not a parish that would suit every clergyman – especially if he had ambitions for advancement. Nor would it be suitable for a wife who valued comfort, or had any social pretensions. But the rural dean had learned many years before that one did not argue with his bishop.

'What was her reason for dismissing you?' he asked.

Alice told them of the unexpected arrival of the new incumbents and their Indian servants, the disturbance of Parson Bushell by Digger and, finally, the arrival of Sailor Smith with Gideon's message that he had gone abroad with the contractor.

'Is Gideon the navvy who came along when you found the Reverend Markham, and who attended his funeral?' the rural dean asked.

'Yes,' Alice confirmed. 'But he's not just a navvy. He's a ganger, in charge of fifty men – and his stepfather is a parson.'

Digger, having ceased his exploration, had been warming himself at the fire. Now he decided to greet the rural dean, who stooped down to stroke him before saying, 'Why have you come here, Alice?'

'I wanted to return what's left of the money you gave me after Parson Markham died, and I had nowhere else to go,' Alice said unhappily. 'I hoped you might be able to recommend me for work as a housemaid somewhere.'

'I can think of no one who would be prepared to take on a housemaid *and* a dog, loyal little chap though I know him to be,' the cleric pointed out.

'I'll try to find a nice home for him,' Alice replied. 'But I couldn't leave him in Treleggan. Mrs Bushell doesn't like dogs. She said she'd have the gardener shoot him if he came into the garden again.'

Making disapproving noises, Beatrice Brimble scooped up Digger. Holding him in her arms, she said, 'What a dreadful threat to make against such a nice little animal!' Looking up at her husband, she said, '*We* could take him in, Harold. You have said many times that the house seems strangely empty since Hannibal died.' To Alice, she explained, 'Hannibal was our little dog. He was no particular breed, but he was very loveable.'

Beatrice's suggestion delighted Alice. 'It would be a great relief if you *could* give Digger a home,' she said. 'Wondering what was going to happen to him has been one of my greatest worries. Parson Markham took him everywhere with him and he couldn't have wished for a more loyal companion.'

'I would say that Arnold Markham was fortunate in having a very loyal housemaid too,' Beatrice said approvingly. 'What do you say, Harold? Do we take in Digger?'

'After a lifetime of devoted service to a servant of the Church, I don't see how I can possibly refuse,' her husband replied. More seriously, he said to Alice, 'I fear that finding a post for you may prove rather more difficult to resolve, young lady. Tell me, are you likely to have this young man . . . this "ganger", calling on you again?'

'I would like to think so,' Alice said honestly. 'He asked me to marry him, but I said we needed to get to know each other a bit better first. Then he was taken out of the country at short notice by the contractor who is responsible for all the work that's being done on the railway in Cornwall. That's what the navvy came to the rectory to tell me. The trouble is, no one really knows when he's likely to be back. The navvy was told to say Gideon would write to me, but if the letter goes to the rectory I don't know what will happen to it.'

'You need have no concerns about that,' Beatrice said firmly. 'I have no doubt my husband will be calling on Reverend and Mrs Bushell in the very near future. He will ensure that if a letter arrives at Treleggan rectory addressed to you it is sent on to us, here. We will forward it to you right away. And you must not worry about anything, Alice, we *will* find a suitable post for you. In the meantime you may stay here and help our servants. You'll find them very pleasant girls. I can also offer you a room of your own. It is quite small, but comfortable enough. Come now, I'll take you there. No doubt Mary can find something dry for you to put on. We will discuss your future in the morning. Would you like Digger to keep you company tonight? It might be a good idea as it will be the first night in a strange house for both of you . . .'

Harold Brimble was able to place Alice in employment far more quickly than he had anticipated. Five days after her arrival at the cleric's Bodmin home, Beatrice Brimble put her on a Plymouth-bound coach to begin a new life. One she could not have imagined only a week before.

She was going to work for a religious order who called themselves 'The Devonport Society of the Sisters of Mercy'. Founded by a devout woman named Priscilla Sellon, the members of the order were popularly known as 'Sellonites'.

Alice was to be employed as a housemaid in their abbey orphanage, leaving them free to carry out the duties to which they had dedicated their lives, namely the care and succour of the poor and sick in the vicinity of the great port of Plymouth.

Somehow the rural dean had learned of the needs of the Sellonites and had written to them, recommending Alice as being a suitable person to serve them. A reply came by return of post. The Reverend Brimble was known to Mother Priscilla. She considered that his recommendation was, in itself, sufficient. She offered Alice employment.

The kindly Beatrice Brimble had supplied Alice with a new uniform in which to commence her employment and the rural dean had another surprise for her before she left Bodmin. He was an executor of the late Arnold Markham's estate and was able to inform her that she had inherited the sum of a hundred pounds from her late employer, a similar sum going to Ivy Deeble.

It was more money than Alice had ever dreamed of possessing, but Beatrice Brimble persuaded her she should not spend it, but would be wise to keep it safely in a bank as an emergency fund, or until such time as she married.

Talk of marriage brought Gideon to the forefront of Alice's thoughts once more. She wished she had an address for him, so she could write and tell him of the dramatic change in her life. But all she had was the promise relayed through Sailor

Smith that 'he would write', and Alice feared the letter might be mislaid before it reached her.

Beatrice Brimble had promised she would send the letter on to her immediately it arrived, but it would need to pass through many hands before reaching her – even if Harriet Bushell kept her promise to put it into the hands of the rural dean.

Alice had been instructed to make her way to St Dunstan's Abbey, the Plymouth base of the religious order, and was told she would be expected. But although the directions given to her by the rural dean were not difficult to follow, it was the first time she had been in a town of such a size. She found the noise, bustle and sheer energy awesome.

When she eventually reached the abbey she was surprised to discover the building was only partly completed. However, the welcome she received from the sisters was both warm and friendly.

The orphanage where Alice would be working was the responsibility of Sister Edith and Alice took to her immediately. Warm-natured and motherly, she possessed a ready smile. After questioning Alice about her life and recent employment, she took Alice to a room that would be her own. She was told she might use the following day to familiarise herself with the building and the routine of the orphans before commencing work.

Alice went to bed that night grateful to the Reverend Brimble for finding work for her in an establishment where she believed she could be very happy.

The next morning, Alice rose early and began exploring the orphanage. It was immediately apparent that the girls in the care of the Sisters of Mercy were far better looked after than the orphaned children she remembered from the unhappy days she had spent among them in the Liskeard workhouse.

The girls themselves performed most of the orphanage chores, but they also received schooling and religious instruction and were taught skills that would enable them to find employment in domestic service when they left the orphanage. Much of

Alice's work would entail helping them to learn those skills and by the end of the first day she was already carrying out such duties.

She thoroughly enjoyed working with the young girls and they seemed to take to her, but the routine of the orphanage was to face an alarming disruption only four days after Alice arrived at the abbey.

She was helping some of the girls to make bread when Sister Edith entered the kitchen accompanied by a tall, rather stern-featured woman who was introduced to her as Sister Frances.

'Alice, we have a problem . . . a very serious problem.' It was the first time Alice had seen Sister Edith without a smile on her face. 'Have you ever nursed sick young children?'

'Not really nursed them,' Alice replied. 'I would sometimes sit with them in the workhouse and give them food and drink when they weren't feeling well – but that was a long time ago.'

'It is better than nothing,' Sister Frances declared, her manner of speech in keeping with her stern expression.

Alice wondered what was coming next, but it was Sister Edith who explained. 'There is an outbreak of cholera in Plymouth, Alice. The first case was diagnosed a fortnight ago. Since then the sisters have been working hard to keep it contained, but two cases have now been reported in the naval orphanage, a short distance from here. It is feared there will be more and we have received an urgent request for assistance. Unfortunately, with all the other cases we have on our hands, only Sister Frances is available. She is one of our most experienced nurses, but will need help. You may refuse, of course. You came to us as a housemaid, not as a nurse.'

Alice had no experience of cholera and the very word was sufficient to provoke a feeling of terror in all who came into contact with the frightening scourge. People who caught cholera died. This was all she knew about the disease. 'I know nothing about actual nursing,' she said uncertainly.

'You will be taught all you need to know by me,' Sister Frances said briskly. She paused for a few moments before giving Alice a direct look. 'These are small children, Alice. The two who are already suffering are aged five and seven. The helpers in the orphanage are afraid to touch them and they have no one else.'

The thought of nursing cholera victims frightened Alice too, but she remembered her own loneliness in the workhouse when she was much older than the two sick little girls. There was also something about this Sister of Mercy that inspired confidence.

'Will the children live?' she asked.

'That is in God's hands, Alice, not ours. All that is certain is that they will die if we do not help them – and probably many others with them.'

Alice thought she was probably quite as afraid of contracting cholera as any of the helpers in the naval orphanage. However, she knew only too well what it was like to have no one who really cared to turn to.

'I'll help, if you tell me what I need to do.'

'Good girl!' Suddenly brisk once more, Sister Frances said, 'Come with me now and we will find out what needs to be done.'

As they walked together to the naval orphanage, the two women talked. It was apparent that Sister Frances already knew a great deal about Alice's background, but she questioned her in some detail about the events which had led to her coming to Plymouth. By the time the orphanage was reached, Sister Frances felt she had a fairly accurate idea of the type of girl who would be working with her.

They found the orphanage in a state of turmoil. Most younger members of the staff had left hurriedly, as had the women who had young families at home. All were terrified lest they too contract cholera.

Their departure had thrown the routine of the orphanage into chaos. The kitchen was virtually at a standstill and it seemed there was nobody in the building capable of re-establishing control.

It took Sister Frances less than an hour to bring some order into the situation. She managed to find sufficient staff to man the kitchen and set them to preparing soup, using whatever food was available. Others were sent to butchers, grocers and

111

greengrocers with a note from the energetic sister, promising that the Sisters of Mercy would ensure that all bills incurred would be paid promptly.

Sister Frances had quickly ascertained that the outbreak of cholera was restricted to a dormitory containing five- to eight-year-old girls. She immediately ordered that those in adjacent dormitories who shared the same toilet facilities should be moved, but kept under close observation. Those actually resident in the dormitory where the infected children lay would remain strictly segregated from the remainder of the orphans.

A crude form of isolation had already been established before the arrival of the two women from the abbey. Such was the fear the outbreak had generated, the children had been locked in since early morning and no one had been near them.

As soon as it was ready, Sister Frances arranged for soup to be taken to the youngest children. Then, taking possession of the keys and accompanied by Alice, she made her way to the stricken dormitory.

The first thing that hit them when they opened the door was the stench. As frightened children ran to greet them, Sister Frances told Alice to open all the windows before turning her attention to the sufferers.

There were now four sick girls, and one was close to death. The bedding of all four was badly soiled and Alice and the sister removed it to the toilets, from where it would be taken to be burned. Fresh linen was found in a cupboard and the beds of the four afflicted children were stripped and remade. As they worked, Sister Frances told Alice it was vitally important that she should wash her hands before and after touching the cholera victims and whenever she had occasion to leave the dormitory.

When the young patients had been made more comfortable, the efficient sister began speaking to the other children. Once her calm and confident manner had quietened their fears, she sent Alice to get clean water and as much flannel cloth as she could find. The two women had brought medicines with them from the abbey and all the girls, regardless of whether or not they were sick, had been dosed by the time soup from the kitchen was brought to them.

As the children ate, Sister Frances and Alice moved the beds of those who were ill to one end of the ward, in order to keep them apart from the others. A fire was also lit, although the windows were kept open.

After a while, Sister Frances left the dormitory to check what was happening in the kitchen and in other parts of the orphanage. She was away for perhaps half an hour. When she returned, she found Alice cuddling Jilly, the most seriously ill of the children, talking soothingly to the little girl, at the same time wiping her face gently with a damp cloth.

Sister Frances looked at the child's gaunt face and half-closed eyes and decided to leave Alice tending her. She knew from bitter past experience that the task would soon come to an end.

'You remain here, Alice. Should you need me I will be in one of the other dormitories.'

An hour later Alice went in search of the sister and found her arranging for fresh linen to be brought from the Sellonite orphanage to replace what had been burned.

Tearfully, Alice said, 'Jilly has died.'

Sister Frances nodded acknowledgement. 'Sadly, it was inevitable. What have you done with her?'

'I wrapped her in a sheet and took her to the orphanage chapel. I didn't know what else to do.'

'You did the right thing, Alice. I will make all the necessary arrangements for her. You return to the abbey, have a cold bath, change your clothes and, if you feel like it, return here. If not, try to find a sister to come in your place.'

Increasingly agitated, Alice finally blurted out, 'I felt so *help-less*. There was absolutely nothing I could do to help her . . . I tried, truly I did.'

Sympathetically, Sister Frances said, 'You gave her what she most needed, Alice. The comfort of feeling loved, something I fear she had experienced little of during her young life.'

Her vision blurred by tears, Alice followed Sister Frances's instructions and ran all the way to the abbey. But she had no intention of leaving the other woman to cope by herself at the naval orphanage. An hour and a half later she returned, accompanied by another sister from the abbey.

For five days, Alice spent most of her time at the naval orphanage, by which time it was evident the worst of the outbreak was over. There had been a number of new cholera cases, but they were much milder than earlier ones. Alice was convinced that the containment of the disease was a direct result of Sister Frances's firm and prompt actions on her arrival at the orphanage.

Whatever the reason, little Jilly was the only child in the orphanage to die as a result of the outbreak of the deadly disease. However, Alice's ordeal was not yet over. Cholera had not been confined to the orphanage. It was raging with frightening ferocity in the narrow streets of the Plymouth slums.

Alice was soon heavily involved there, helping those affected by this far more serious outbreak, and the Sisters of Mercy quickly realised that she could be relied upon to work without supervision. She performed valuable work for almost three weeks, by which time the worst of the epidemic was over and the town's own resources were able to cope.

A housemaid's duties seemed mundane to Alice after the responsibility she had shouldered in preceding weeks – but such a dull routine was not destined to last for very long . . .

10

I

It was mid-October when the ship carrying Gideon Davey and Ranald MacAllen reached Constantinople, the great Turkish city which was the gateway to the Black Sea, where the war-ravaged Russian peninsula of the Crimea was located.

Obliged to change ships at Constantinople, Gideon and Ranald were escorted to nearby Scutari in order to take passage on a ship to the Crimean port of Balaklava. Captured by British and French troops a few weeks earlier, Balaklava was the port through which troops and supplies reached the British army.

The two newly arrived men had been told that the ship on which they were to embark was waiting at one of the quays. However, when they reached the quay they found the ship was now anchored offshore, its place taken by a steamship, newly arrived from Balaklava with a tragic cargo of ill and dying British soldiers. Gideon and Ranald were horrified by the sight of hundreds of ragged and filthy military men resting on the quay-side while others were being stretchered from the ship. There was an air of utter defeat among them.

'Have we just lost a battle in the Crimea?' Gideon's question was put to Ranald, but his words were overheard by a well-dressed, pipe-smoking Englishman standing nearby.

Taking the pipe from his mouth, he said, 'No, gentlemen, there has been no battle, and if the Russians have sense enough to do nothing they will win the war without firing another shot. The soldiers you see here are the casualties of inept leadership. A commander-in-chief who has never led men into battle, and a cavalry commander who last saw action a quarter of a century ago – fighting on the side of the country that is now our enemy. Add to this a Light Brigade commander who abandons his men to mud and discomfort before dinner each day in order to return to the safety and luxury of his private yacht, and there you have it.'

Gideon and Ranald were taken aback by the stranger's outspoken criticism of the army's military leaders. He, in turn, seemed amused by their startled expressions. After drawing on his pipe to keep it alight, he removed it from his mouth once more. 'Allow me to introduce myself. William Turnbull – described by my London newspaper as their "war correspondent". But you are wearing no uniforms, gentlemen. Do you have business with the war?'

'I am Ranald MacAllen and this is Gideon Davey. We are in the employ of Robert Petrie . . .'

Before Ranald could continue, Turnbull interjected excitedly, 'Petrie! The railway constructor? So it's true then? I recently telegraphed to London that rumours are rife here of a railway line to be built from Balaklava for the benefit of the soldiers investing Sebastopol. Poor wretches like these . . .' He made an expansive gesture that encompassed the sick soldiers all about them on the quayside. 'Look at them, poor souls. The victims of such colossal military incompetence that the great Duke of Wellington must be rotating in his tomb. But where are the rest of your men, good sirs? Are they already on their way to the Crimea?'

'No, Mister Turnbull,' Gideon replied. 'Ranald is the engineer and surveyor in charge of the project, and I will be in charge of the navvies. We are on our way to find out what will be needed. Mister Petrie would have been with us, but he suffered a nasty accident in France.'

William Turnbull groaned. 'I suppose that means work will not begin until the weather improves. In that case it's unlikely to be completed before *next* winter. I fear our army will have ceased to exist long before then.'

'Then we must ensure the railway's completed as quickly as possible,' Gideon commented. 'It shouldn't take too long once the men and materials are here.'

'I fear you are unduly optimistic,' Turnbull replied. 'You have not seen the state of the countryside in the Crimea – but are you on your way to Balaklava today?'

'That's right,' Ranald said, 'travelling on the *Mermaid*. We were told we would find it moored alongside the quay, but it seems it's been moved.'

Pointing to a ship anchored some distance offshore, Turnbull said, 'That's the *Mermaid*, out there. I'm taking passage on it myself. I have a boat ordered for two o'clock. Bring yourselves and your possessions here and we'll go out to the ship together – but sit on your baggage while you're waiting. It will disappear if you turn your back for more than a second.'

Gideon and Ranald had intended taking a walk round Scutari, where there were two hospitals housing British casualties from the Crimea. One, a huge, ugly white building, was a former barracks, the other was the general hospital. Both were filled to overflowing.

However, they soon discovered that William Turnbull's warning was fully justified. After twice succeeding in recovering items of baggage stolen by opportunist Turkish thieves who loitered around the quayside, the two men decided to remain with their belongings until it was time to board the *Mermaid*.

Turnbull arrived accompanied by a small army of Turkish porters who carried his baggage and a great deal of photographic equipment.

The correspondent was in a jovial mood. He had been entertained to lunch by the medical officer in charge of one of the

Scutari hospitals. The surgeon had informed him that a British nurse by the name of Florence Nightingale was being sent to Scutari to take responsibility for the care of the increasing number of British casualties. With Miss Nightingale would be approximately forty trained nurses, a number of them members of religious orders.

'Don't they already have nurses at the hospitals?' Gideon asked the question in all innocence.

Turnbull's lunch had been alcoholic enough to satisfy the needs of the most Bacchanalian of journalists, making him even more loose-tongued than usual.

'There isn't a single skilled nurse in the whole of Scutari and nursing orderlies are detailed from those men who have recovered sufficiently to carry out the work. Most would murder a dying man if they so much as glimpsed the glint of gold in his purse. As for the *women* who call themselves nurses . . . some are humane enough – when they're sober – but most will rob a man if they believe he's dying, or will share his blanket – for a price – if he has strength enough to want them. Either way they'll have his money before he leaves Scutari. Perhaps I should say *if* he leaves Scutari. The hospitals are a disgrace, with a mortality rate in excess of fifty per cent. Most wounded men would rather take their chances in a mud-floored tent hospital in the Crimea than be sent here.'

Observing the scepticism of his listeners, he said, 'I see you do not believe me, gentlemen. If the wind is blowing offshore as we set sail you will change your mind. The stench is enough to turn the stomach of the strongest man.'

The wind was not blowing from the land when the *Mermaid* left Scutari and Gideon and Ranald remained sceptical of the war correspondent's stories until the moment the ship eased into a berth in the small and overcrowded harbour at Balaklava.

It was early in the morning and the weather was cold with a needle-sharp drizzle carried on the wind, yet hundreds of ill and maimed soldiers were huddled together without shelter, in miserable groups on the dockside. They possessed only a scattering of shared blankets to protect them from the elements. Others lay on rows of stretchers and, although some had a canopy of sagging, water-filled tarpaulins over them, many were lying in the open.

Gideon expressed surprise to see women and a few children among the troops. Turnbull had taken to his cabin when the weather had deteriorated en route to Balaklava and emerged only when the ship sailed into more sheltered waters. Now he explained these were wives and children of soldiers, a number of whom were allowed to accompany their men, even to the scene of an impending battle.

'There's nothing unusual in that, surely,' Turnbull added. 'Don't your navvies take their women along when they're building a railway?'

Gideon had to admit that they did, but said, 'There will be no women out here with our navvies. They'll be left at home in safety. The men will work all the harder in order to return to them as quickly as they can.'

Dismissing the matter, Turnbull said, 'You'd better both come ashore with me. I'll find rooms for you in the house where I'm staying and show you round Balaklava.'

'It's a kind offer,' Ranald replied, 'but we have letters of introduction to General Sir John Burgoyne, the army's chief engineer. We need to report our arrival to him right away.'

The previous evening Gideon and Ranald had agreed it might be best if they distanced themselves from William Turnbull, at least for the first few days of their stay in the Crimea. The war correspondent had an abrasive manner and was scathing in his comments about the officers who led the British army. He would undoubtedly have antagonised the very men whose co-operation would be so important to their mission. It would not be wise to be thought of as being too friendly with him.

Turnbull made a derisive sound in his throat at the mention of the chief engineer. 'Burgoyne! He hasn't seen any fighting since he was in Wellington's army in the Spanish peninsula, nigh on fifty years ago. Get him involved and the last British soldier will be on his deathbed long before a railway gets

119

started. Had Lord Raglan and the French commander-in-chief not heeded his advice to wait, Sebastopol would have been ours within days of our troops coming ashore! Mind you, Raglan needed little persuading. He has little stomach for war. I don't suppose we should blame either of them. Men of their age should be at home, sitting in a chair with a blanket wrapped about their knees, not out here in the mud and blood of war trying to out-think an enemy.'

'How old is Burgoyne?' Gideon asked, reluctant to believe that someone who had fought in the Napoleonic Wars, fifty years before, should hold such an important army post in the Crimea.

Satisfied that his revelation had startled Gideon, Turnbull replied, 'General Sir John Burgoyne is seventy-two years old. Mind you, Raglan himself is not much younger. All the British generals are much of an age, each trying to recapture the glories of generals he followed in his youth. Unfortunately, this is a very different war, in a different land – and not one of them has a fraction of the talent of the ones who led them then. Men like Moore and Wellington.'

As Turnbull went to his cabin to prepare to go ashore, Gideon and Ranald exchanged glances. They both felt their decision to distance themselves from the correspondent had been the right one. His outspoken views were not likely to have endeared him to senior officers of the British army.

II

The two railroad builders were fortunate enough to find General Sir John Burgoyne on the Balaklava quayside. He had come to bid farewell to one of his closest aides, invalided home after suffering a severe bout of cholera, the disease that was rapidly bringing the British army to its knees.

Although quite as old as Turnbull had declared him to be, the general was not lacking in energy. Within an hour he had found accommodation for the two men in a house in Balaklava; set a time for a meeting that afternoon with some of the army's own engineers; and arranged for Gideon and Ranald to be taken to the siege lines around Sebastopol, travelling there along a possible route for the proposed railway.

The meeting with the senior engineers was a depressing affair. The majority were of the opinion that construction work on a railway would not be possible until spring, when the weather improved.

'The land between here and the siege lines has been turned into a quagmire,' one of them explained. 'Horses sink into it above their hocks. No men can work in such conditions.'

'You are underestimating my navvies,' Gideon said confidently. 'They've built railways in every continent in the world and have never yet allowed bad weather to stop them.'

'Then it's a pity we cannot put them in uniform and allow them to fight the Russians for us,' General Burgoyne commented. 'At the moment weather and disease are the only victors in this war. But that's enough talking. I've had one of my sappers draw a map for you of the land between here and our lines around Sebastopol. He has included contour lines as far as has been possible in the time at his disposal. He doesn't guarantee absolute accuracy, but the map is a damned sight better than any that were available to us when we landed. It'll give you something to study this evening. I have arranged for an escort to call for you at your lodgings in the morning. They will be at your disposal for the whole of the time you are here, to ensure you do not stray too close to the Russians. You will do well to follow their advice. Thank you, gentlemen, that will be all.'

The meeting broke up with the gloomy army engineers still expressing their doubts about the practicality of trying to build a railway line in the present weather conditions.

The next day, 25 October 1854, was one Gideon would always remember, along with the rest of the world. It was to prove that William Turnbull, although a thorn in the flesh of the

War Office, was justified in his assessment of the incompetence of the senile generals who led the British army in the Crimea.

The escort provided by General Burgoyne comprised a dragoon sergeant and six troopers, with two spare horses for Gideon and Ranald. Unlike the infantrymen, who were the only soldiers Gideon had so far seen, these cavalrymen were smart and their equipment was well looked after, even though their horses appeared in need of a good feed.

Their leader introduced himself as Sergeant Philpott, saying, 'We've been detailed as your escort, sirs. I believe you want to check the country between here and Sebastopol. With all due respect, I suggest you put it off for a day or two. When we left headquarters reports were coming in of Russian troop movements to the north-east. There's a strong possibility they intend launching an attack against our redoubts up in the hills.'

'No doubt there's always a possibility of an enemy attack,' Ranald said firmly. 'There's a war going on. But we have a task to perform – although, from what I've seen of our army so far, we might already be too late.'

'I wouldn't argue with that, sir,' the sergeant said. 'Very well, we'll do our best to ensure your work is carried out in safety. I was just warning you, that's all.'

'Thank you, sergeant.' Producing the map provided by General Burgoyne, Ranald handed it to Philpott. 'I've drawn in the route that seems the best to follow, but we're likely to change our minds on occasions along the way, if either Gideon or myself thinks it advisable.'

After studying the map, the sergeant handed it back to Ranald. The route the railway was likely to take through Balaklava itself was both bold and direct, involving the destruction of a great many houses. Gideon could not resist a secret smile when he realised the proposed route would pass through the garden belonging to the house where William Turnbull lodged. Indeed, the line would be little more than an arm's length from the house itself.

It was apparent that before the arrival of the British army,

Balaklava had been a very pretty little town, popular as a holiday resort for wealthy families from the interior. However, tens of thousands of men, horses and supply wagons had devastated the area to such an extent that in its present state it would be quite unrecognisable to former visitors.

The only land route out of Balaklava was through a pass in the surrounding hills. It was the channelling along this pass of all traffic pertaining to the invading army, together with the onset of winter, that had resulted in the area's becoming a near-impassable quagmire.

It was so bad that Ranald was seriously concerned about the prospects for a railway line. He expressed his misgivings to Gideon as their horses struggled through the tenaciously clinging mud.

'This terrain is as bad as any I've had to survey for a railway, Gideon, but it's the only route we can take. We can't go over the hills and tunnelling through them would take more time and resources than we have at our disposal. What's your opinion?'

'I think we need to find a quarry somewhere in these hills and bring broken rock in to give us a base on which to lay the rails,' Gideon replied. 'It shouldn't be too difficult. Work can start on that, using local labour, while we wait for the navvies and equipment to arrive. There should be enough rock up there. If there isn't, we can use rocks and pebbles from the shore.'

'That won't do away with the mud,' Ranald persisted. 'The work is going to be well-nigh impossible for your men.'

'If the army can fight a war in such conditions they won't stop my navvies from working,' Gideon declared, adding, 'You decide on the route, Ranald, we'll build you a railway.'

While they were talking both men had heard a sound like a continuous rumbling of thunder, and now they became aware of the anxious glances exchanged between the sergeant in charge of their escort and his men.

'What is it?' Gideon called to the soldiers.

'It's artillery, sir – lots of it. If you listen you can hear musket fire too now.'

As they neared the end of the pass they could also hear the shouting of men – many men – in addition to the firing.

'Something serious is happening,' said the sergeant. 'I think I should go ahead and find out what it is.'

'We'll all go,' Ranald said. 'With only one route for the railway there's not much surveying to be done until we are clear of the pass. Come on, Gideon, let's see what's going on.'

He dug his heels into the flanks of his horse, but it made little difference to the animal's speed. It was in poor condition and with the mud so deep progress was laboriously slow.

When the small party eventually cleared the pass they entered a wide shallow valley running from left to right in front of them. A small hill to the right of the riders obstructed their view of much of the valley. It was a natural sentinel guarding the entrance to the Balaklava pass and on this mound a few hundred Argyll and Sutherland Highlanders were forming a loose double line, their backs towards Gideon, Ranald and their escort.

'There *is* something going on. I'm off to have a look.' The going was easier here, and even as he was speaking the sergeant was urging his horse on.

He had not gone far when he pulled his mount to an abrupt halt. Following close behind him, Gideon heard him exclaim, 'Good God!'

Reaching the sergeant, Gideon saw the cause of his astonishment.

It could now be seen that the long valley was far wider than it had first seemed. What Gideon had thought was the far boundary was in fact a range of low hills, a 'causeway', running the length of the valley and dividing it in two.

On this causeway a mass of Russian and British cavalrymen were locked in ferocious battle. Closer at hand, a great number of Turkish infantrymen were fleeing the scene, running towards the pass that led to Balaklava. The shouting came from the cavalrymen of both sides as their formations formed and broke, and formed again, all the time thrusting and hacking at each other in a tangled frenzy of horses and men.

'That's our regiment up there, with the Heavy Brigade – we

should be with 'em.' The sergeant spoke as though talking to himself. Suddenly turning to his companions, he said, 'Jenkins, Skinner, Rawlins and Cooper, come with me. We're going to rejoin the regiment. Ellis and Farrell, escort these two civilians to Lord Raglan's headquarters, on the Heights.'

Without waiting for a reply, the sergeant wheeled his horse. Followed by the four chosen troopers he galloped towards the scene of the battle, leaving behind the other two disgruntled cavalrymen.

'Come on, we'll get you to headquarters – then come back here, if it's not too late by then.'

With one trooper leading and the other bringing up the rear, the depleted party made all speed possible. Soon they left behind the churned-up mud of the valley and began following a narrow track that led up into the hills at the opposite end of the valley from where the battle was taking place.

Suddenly, the trooper bringing up the rear of the party called, 'Look! There's the Light Brigade – and the Russians, thousands of 'em!'

They could now see the half of the valley that had been hidden from view by the causeway heights. It became perfectly obvious to Gideon that what they had already seen was no more than a fierce skirmish. It was here, unseen at first by the Heavy Brigade of cavalry, that the most serious fighting would take place.

The cavalryman with them had not exaggerated when he said there were thousands of Russians. Gideon estimated there must have been *tens* of thousands. They were drawn up in formation on the slopes of hills at the far end of the valley with cavalry and artillery in front of them – and still more guns and infantry were taking up positions on the hills on the far side of the valley, and on the slopes of the causeway.

The British cavalry, referred to by one of the two-man escort as 'the Light Brigade', were formed up at the near end of the valley, almost directly beneath the path on which Gideon and Ranald sat their horses.

There were pitifully few of them in comparison to the vast

Russian army, a valley's length away, and a heated discussion appeared to be taking place among a small party of the British cavalry officers. As Gideon and the others watched, one of the officers began gesticulating wildly in the direction of the Russian army.

Moments later the argument was over. Orders were shouted, and the cavalrymen kneed their horses into formation, first into two lines, then, at another shouted order, into three.

One of the two cavalrymen beside Gideon said, in utter disbelief, 'Surely they're not going to attack in the face of all those Russian guns?'

'There are almost as many on the hills on either side,' said his colleague grimly. 'To make a charge along the valley would be suicide.'

As if in defiance of his words the notes of a bugle reached the watching men and the three lines of light cavalry began moving forward in near-perfect formation.

An eerie hush suddenly descended upon the whole battlefield, as friend and foe alike paused in what they were doing, disbelief momentarily taking the place of an urge to kill.

For a short time the only sound to be heard was the jingling of the harness and accoutrements of the men and horses of the Light Brigade, but the silence did not last for long. The British cavalry was advancing at a steady trot when Russian guns lined up at the far end of the valley thundered into life.

As shot landed among them, many gaps appeared in the ranks of the Light Brigade, but in response to the shouts of their officers the lines closed up, leaving dead and dying men and horses in their wake, and the cavalry continued its advance at the same, seemingly leisurely pace.

Now they were within range of the Russians holding the hills on either side of the valley and a withering fire was poured into their ranks from the muskets of the infantry and the guns of the Russian artillery. Fired upon from three sides and being mowed down in increasing numbers, the Light Brigade quickened their pace, the trot becoming a canter.

Meanwhile, the Heavy Brigade of cavalry, having decisively

126

routed superior numbers of Russian cavalry and becoming aware of what was happening below the causeway, hurriedly re-formed and began to ride in support of the Light Brigade.

Fortunately for the cavalrymen involved, they were brought to a halt once they came within range of the Russian guns and began taking casualties. Reluctantly, the eminently sensible order was given for them to retire. The Light Brigade was once more on its own – but it was not to be left entirely at the mercy of the enemy without some support.

Appearing as if from nowhere, a body of mounted men that had, until now, been hidden from view by a fold of the hills swept into action, erupting from the heights on which Gideon, Ranald and the two troopers were viewing the battle.

'It's the French – the African Chasseurs,' one of the troopers said excitedly.

'But . . . what can they do?' Gideon asked.

For a moment it seemed they would follow the same bloody route as the Light Brigade, but then they changed direction and, in no discernible formation, swept along the side of the valley from which the most withering musket fire had poured into the ranks of the Light Brigade.

The Chasseurs d'Afrique could do nothing to save their allies who had already passed along the valley pursuing their reckless attack, but, driving Russian infantry and artillery before them on the hillside, they ensured that any survivors would have less to contend with when they retired.

By now the Light Brigade had broken into a gallop and, suddenly, all semblance of a formal formation was lost. Men and horses were crashing to the ground in distressing numbers, yet those who remained charged on through the smoke, bullets and shells, their cheers and shouts of defiance rising above the cries of wounded men and horses.

After a last, point-blank salvo that virtually wiped out those in the van of the charge, the guns fell silent as the remainder of the Light Brigade fell upon the artillerymen, stabbing and slashing in vengeful fury at the soldiers who had slaughtered and maimed so many of their comrades.

Hacking their way through the line of guns, the surviving British cavalrymen found themselves confronted by an over-whelming force of Russian cavalry, including the much vaunted Cossacks.

Pausing only long enough to form up and close ranks – wherever it was now possible – the survivors of the Light Brigade charged the Russian cavalry and, amazingly, cut their way through them!

The Russian artillery had been silenced, many of their gunners cut down, and the Russian cavalry taken by surprise, but the odds against the Light Brigade were far too great for them to stand any chance of success in a pitched battle. Calling for the survivors to rally to them, the few remaining British officers turned their horses and prepared to sell their lives dearly as they retired from the scene of the furious and confusing mêlée.

Attacked by Cossack cavalry, fired upon by riflemen and artillery from the central causeway, the retreat seemed painfully slow to the watchers on the Heights.

Although a few cavalrymen still sat their horses, many were on foot, some leading mounts with disabled colleagues in the saddle. Others tried to drag injured animals after them.

The two cavalrymen with Gideon and Ranald could contain themselves no longer. Kneeing their horses into motion, they rode down from the Heights as fast as the slope would allow, and the two railwaymen followed in their wake.

The cavalrymen knew some of the survivors and within minutes had given up their mounts to wounded and exhausted men. Meanwhile, Gideon and Ranald rounded up four horses between them and passed them on to cavalrymen who were fleeing on foot.

By now columns of British infantrymen were streaming down from the Heights and taking up formation in the valley. Lord Raglan, the British commander-in-chief, and his staff had also descended from their headquarters on the Heights, from where the battle had been so disastrously conducted.

Gideon and Ranald decided they could not continue with their survey that day. Retracing the route to Balaklava, they made their way to their lodgings in the town. Neither man felt inclined to talk of what they had seen. The sight of so many dead and wounded men and horses would remain with them for ever.

Their mood was not lightened when the sergeant of their escort found them later that evening. He would be there to guide them on their route in the morning, but there would be replacements for two members of the escort. They had reached their comrades of the Heavy Brigade in time to take part in the final stages of the battle. One had been killed, the other badly hurt.

The sergeant also brought news that of the 658 men of the Light Brigade who had taken part in the heroic but disastrous cavalry charge, only 195 had returned. In addition, some 500 horses had been killed.

The result of the disastrous British charge meant that Lord Raglan's army no longer had a brigade of light cavalry.

11

I

Alice spent most of her time in the abbey orphanage working with the youngest children – work she loved. It also allowed her the time to think, and many of her thoughts were of Gideon. She worried about him now because she had learned something of the war going on in the Crimea and realised this must be the reason he was there, although she could not imagine what his part in such a conflict could possibly be.

Thinking about him so often, she wondered whether his thoughts were of her. She also speculated how different her life might be had she been more responsive to him after their day out together. But she told herself, firmly, that was in the past. Nothing could be changed now.

She consoled herself with the thought that whatever she might have said or done would not have prevented him from leaving England for the Crimea. However, it would have been far more settling had there been an understanding between them before he left. There was the question of his letter, too. She wished it had reached her before she left the Treleggan rectory so abruptly.

The thought of Gideon was a constant concern and his

promised letter assumed ever greater importance with each passing day. Her great fear was that it would arrive and be thrown away or burned by Harriet Bushell, despite Harold Brimble's assurance to her.

She could not express such thoughts to the rural dean, but she did write a letter to him and Beatrice Brimble, thanking them for their great kindness to her, and saying how happy she was to be working for the Sisters of Mercy. Enquiring after Digger, she – almost casually – asked whether the rural dean had remembered to mention to the Reverend Bushell that a letter should be arriving at the Treleggan rectory, addressed to her.

She received a reply from Beatrice Brimble a mere forty-eight hours later. Although nothing had been received from Gideon, the Reverend Brimble had given firm instructions to Parson and Harriet Bushell that any letter received at the rectory addressed to Alice should be forwarded to him. He promised it would be sent on to her immediately.

Harold Brimble had also heard from Priscilla Sellon. She had told him of Alice's invaluable work during the recent cholera outbreak. Beatrice declared she and her husband were delighted that their faith in Alice had been fully justified.

The letter went on to say that on his visit to the Treleggan rectory, Harold Brimble had been told by Harriet Bushell that her ultimatum to Alice had been no more than a threat, intended to put Alice 'firmly in her place'. What was more, the new rector's wife had suggested that if Alice wished to return to Treleggan and adapt to the ways of the new incumbent and herself, she would be reinstated.

With only a hint of unchristian malice, Beatrice Brimble suggested that the change of heart may well have been prompted by the fact that the Treleggan villagers seemed to have taken a dislike to the wife of their new rector with her colonial ways and Indian servants. They were doing little to help her settle into the isolated moorland village.

Beatrice added that her husband had made it clear to Harriet Bushell that Alice's value was recognised within the Sellonite community and he doubted whether she would consider a return to work at the rectory.

He had also made it clear that *he* had never informed Alice of Emmanuel Bushell's appointment and imminent arrival at Treleggan. In view of Henry Stanbeare's well-known attitude towards Alice over the years, the Reverend Brimble suggested there was a strong possibility the churchwarden had 'forgotten' to tell her they were coming.

Beatrice Brimble's letter gave Alice a degree of satisfaction about the unfair manner of her dismissal from the Treleggan rectory. She was also more confident now that a letter from Gideon *would* be sent on to her, but she was increasingly concerned that it had not yet arrived.

In fact, the letter from Gideon reached St Dunstan's Abbey only five days later, but by then Alice had been gone from Plymouth for twenty-four hours.

Alice's departure from the abbey was even more sudden than her departure from Treleggan rectory. She had completed her duties in the orphanage late in the evening and was in her room, preparing for bed, when there came a knock on her door. When she opened it, Sister Frances was outside.

'I am sorry to disturb you so late in the day, Alice,' Sister Frances apologised, 'but I am here on a matter of great importance . . . may I come in?'

Once inside, Sister Frances noted with silent approval the tidy state of the room, but she wasted no time before disclosing the reason for her unprecedented visit.

'Alice, the Reverend Mother has had an urgent call from London. There is a desperate need for nurses to go to Scutari, in Turkey, to tend our soldiers in the hospital there. Eight Sisters of Mercy have been chosen to join Miss Nightingale and a party of about thirty nurses travelling there. We are leaving for London tomorrow and setting off for Scutari the following day. I know you are not a sister, but the work you did at the naval orphanage was invaluable. I have suggested that you accompany us, and the Reverend Mother agrees. So, if you are willing, I would like you to travel with us.'

Alice had never heard of Miss Nightingale, but she realised she was being asked to go to a hospital for wounded soldiers in the company of women who had a knowledge of nursing that was far superior to her own.

'But . . . I know nothing about nursing wounded soldiers,' she protested.

'None of us do, Alice, but from what I have been told the vast majority of the soldiers – and their wives and children – are also suffering from cholera, something you have had more experience of than most, and have proved you are well able to cope with.'

'How long will we be there?' Alice's thoughts were of Gideon. She had not connected Scutari with the war in the Crimea.

'I don't know how long *we* will be there,' Sister Frances replied, 'but you will be going as a free agent. You may return whenever you wish – although I fear there will be much to do. Is there a particular date when you want to be back in England?'

'I don't know,' Alice replied unhappily. 'I am waiting for a letter from . . . from someone who is in the Crimea.'

'But Scutari is the hospital for the Crimea,' Sister Frances explained. 'The soldiers we will be treating are those who have been fighting in that war. If this "someone" of whom you talk is there you are likely to meet up with him.'

'Do you really think so?' Alice became suddenly excited. Sister Frances's words had put an entirely different complexion on the issue.

'If we don't meet up with him right away we can certainly get word to him that you are at Scutari . . . if that is what you wish?'

'I do!' Feeling she should explain in more detail, she added, 'His name is Gideon. He . . . he has asked me to marry him.'

'Well! It could be a very happy reunion . . . but I must remind you we will be there to fulfil a most exacting task.'

'Of course,' Alice said. 'It wouldn't stop me from working.

133

It's just . . . it would be wonderful to know he was close.'

'Then I may tell the Reverend Mother you will accompany us?'

Alice nodded. 'Yes . . . but I would like you to teach me more about nursing on the way there.'

'Good girl! I will go and inform the Reverend Mother and leave you to pack your things. As for nursing instruction . . . I will teach you myself. By the time we arrive at Scutari you will be as skilled as anyone else in the party.'

II

The following day Alice caught the train to London with Sister Frances and two other sisters. Five more Sisters of Mercy were either already in London, or on their way to the capital from various establishments around the country.

That night was spent in St Saviour's Hospital, home of another religious order. The following day was a Sunday and after the sisters had attended a morning church service, they and Alice, together with a number of secular nurses, were taken to the Belgravia home of the Secretary for War. Here, the Secretary himself spoke to them of the importance of the task they had undertaken and gave them a stern lecture about their conduct while they were at Scutari. The sisters were not to attempt to convert soldiers to their religion and secular nurses were warned to behave with modesty and decorum at all times.

Alice was surprised that the latter warning was deemed necessary, but, as Sister Frances explained to her afterwards, so-called 'nurses' employed in military hospitals had earned an unsavoury reputation for promiscuity and drunkenness.

It was only now that Alice realised the full importance of the task that lay ahead, to which she had agreed without giving it any serious thought beyond the exciting hope that it could lead to an early reunion with Gideon.

Had it not been for this, she might have been inclined to

change her mind. It seemed she and the others would be subject to an extremely strict regime.

The remainder of that day was spent in a whirlwind of documentation, instruction and preparation for the long journey that lay ahead of the women, and the following morning they rose long before dawn.

By 6 a.m. Alice and the sisters were on their way to London Bridge station. Here they met up with others of the party. In all, a total of more than forty women would be travelling together to Scutari.

By breakfast time they were in Folkestone, where they boarded a ferry to Boulogne. Once in France they journeyed by train to Paris where they were joined by more nuns who were making up the party heading for Scutari.

On the journey through France, Alice was unknowingly following the route taken by Gideon a few weeks before. In fact, a convalescing Robert Petrie was in the party from the British Embassy who came to the Paris railway station to see them depart on their way to the southern French port of Marseille.

Travelling through France was probably the most exciting time of Alice's young life. It was not only the thrill of seeing new places and people, but also the experience of being treated like royalty by the ordinary French people once they realised who the English women were, and where they were going. The French army in the Crimea was almost twice as large as the British contingent and people here seemed much more aware of what was happening there.

However, although Alice had been selected by Sister Frances because of her experience in tending cholera victims, she soon realised that others in the party did not accept her as a nurse.

Florence Nightingale was going to the Crimean theatre of war on a personal crusade. It was her declared intention to give the nursing profession a status it had never known before. She would make it respectable. Each of the qualified nurses who formed her party, religious or secular, had been personally interviewed and recruited by Miss Nightingale herself. All were

135

drawn from upper-class backgrounds and some were titled. It was clear they regarded Alice as a servant, and no more than that. She was not invited to any of the civic receptions held for them on their journey through France.

This did not unduly upset Alice because the Sisters of Mercy and a small number of the nuns who were travelling with them declined to attend such functions. However, one of the secular nurses, Josephine Wotton – the Honourable Josephine Wotton – increasingly delegated menial tasks to Alice throughout the long journey, in the apparent belief that Alice was travelling with them as a servant to whoever required her services. It caused Alice to wonder exactly what her role would be when they eventually arrived at their destination.

After four days of travelling through France, the party arrived at the busy Mediterranean port of Marseille. Here, as in so many other towns, they were cheered and applauded as they made their way to the *Tigress*, the ship that was to take them to Scutari, where their work would commence.

The welcome extended by the French people towards the British nurses was not echoed by the Mediterranean weather, although it gave no indication of what was to come when they set sail on the next stage of their long journey.

It was considerably warmer than it had been when they set off from England. Those who had never been to sea before were talking of spending the next few days basking in the sunshine on the deck of the *Tigress*.

Alice's only experience of the sea had been on the great paddle-steamer which had taken her and Gideon to Falmouth from the small harbour of Charlestown and she too was looking forward to this voyage.

Thoughts of that earlier occasion brought Gideon to mind more vividly than usual and she boarded the vessel wondering how long it would be before they met once more. She had no idea how close Scutari and the Crimea were to each other but did not doubt they would meet again soon after she and the nurses arrived at their destination.

The voyage on which Alice was now embarked had little in

common with the happy and comfortable trip she had enjoyed between Charlestown and Falmouth. That had been a brief outing on a large vessel in sheltered coastal waters.

Here, in the Mediterranean, she and the others were on the open sea and the *Tigress* was a mail packet, a ship designed for the rapid delivery of mail, its narrow lines sacrificing comfort for speed.

The weather was quite pleasant when the ship left Marseille, but it was soon apparent that the vessel tended to attack the sea, even in relatively calm conditions. Dipping into a wave, instead of rising again immediately it would dip a second time, before shaking free of the water with a peculiar, juddering movement that was both uncomfortable and disconcerting.

By the time the *Tigress* had been at sea for four hours, all the nurses had retired to the bunks which had been erected for them in makeshift cubicles in the forward hold. It was airless, cramped and dark in the hold, but lying down here was preferable to trying to maintain a balance anywhere else on the ill-designed vessel.

There was a brief respite for the seasick women when the ship entered the harbour at Valetta, ancient capital of the island of Malta, the erratic movement of the steamship ceasing as it nosed its way into the crowded harbour.

Overjoyed to be free of the ship's nauseous motion for a while, the nurses would have been happy to spend their brief respite cleaning up and resting after such an uncomfortable beginning to their sea voyage. Instead, they were ushered ashore and virtually marched round Valetta at a light infantry pace on a prearranged sightseeing tour, intended to show them off to the local populace.

Alice was invited to go with the nurses on their swift sightseeing tour, but as they were preparing to go ashore a boatload of women and children arrived alongside the anchored *Tigress*. They were army families who had been accompanying their soldier husbands and fathers to the Crimea when there had been an outbreak of measles on board the troopship. Those affected were landed at Malta and left behind in an island hospital. Now, fully recovered, they were anxious to resume their journey to join their men in the Crimea.

The families were allocated space in the forward hold, despite the protest of Sister Frances that it needed to be thoroughly cleaned before convalescent women and children could be accommodated in the already crowded space.

Always eager to please the sister who had arranged for her to travel to the Crimea, Alice said, 'I'll stay on board and begin cleaning up, Sister Frances. No doubt some of the women who've just come on board will help.'

She set about the task immediately, but not all the soldiers' wives were happy to take on the task of cleaning up the mess made by the seasick nurses. Nevertheless, five of them joined Alice and she found it something of a relief to be working among women from a similar walk of life to her own.

Among the new arrivals on board was a young woman only a couple of years older than Alice. Travelling with her was a young daughter of three years of age.

Placing the child on the bunk that mother and daughter would share for the journey to the Crimea, the woman took up a mop. Working alongside Alice, she introduced herself as Gwen Dymond and her daughter as Verity, adding, 'I heard you speaking to the nun just now. You have the same accent as my husband. Unless I'm very much mistaken you're from Cornwall.'

'That's right,' Alice agreed, 'I'm from a little moorland village called Treleggan.' Talking of Treleggan stirred up strange feelings inside Alice. It was a name she would happily have banished from her memory, but knew she never would. Hurriedly, she said, 'Where is your husband from?'

'His parents have a farm on Bodmin Moor – isn't Treleggan there too? I don't come from Cornwall myself, but I'm sure I've heard Jack – he's my husband – mention it.'

'It's very near! In fact, I think I've met your husband's father. His farm is actually in Treleggan parish and he came to the rectory where I worked once or twice.'

Jeremiah Dymond owned a fairly large farm on the parish

138

boundary to the north of Treleggan. Parson Markham had liked him and the two men had a great respect for one another. However, Alice had no wish to talk of Treleggan. It belonged to an unhappy past. 'Is your husband in the Cornish regiment?' she asked. 'I didn't realise they were in the Crimea.'

'As far as I know, they're not,' Gwen replied. 'My Jack's a gunner – with the artillery. We were with the garrison on Gibraltar when he was ordered to the Crimea. We were on the troopship for almost a week before leaving and some of the children began to get ill with the measles, but we sailed anyway. It broke my heart when Verity caught the measles too and the doctor on board decided that all the sick children and their mothers should be put ashore, here in Malta, while Jack and the other husbands went on without us. I keep praying that he's all right and that we catch up with him before he goes into action.'

'I'm sure you will,' Alice said reassuringly. 'But it looks as though Verity has fallen asleep. You go and make her comfortable. Take some bedding from my bunk – that's it in the corner, over there.' She pointed to a bunk in the narrowest section of the hold. 'When I've finished here I'll go and collect bedding for you and the others.'

Seeing the small girl lying asleep on the bare boards of the bunk had brought back unhappy memories of Jilly, the small girl who had died from cholera in her arms, and of the many other children in the naval orphanage.

The rules had dictated that as soon as they rose in the morning, the children had to roll up their wafer-thin mattresses and place them with carefully folded blankets at the end of each bed. Alice had always believed that of all the many regulations in the orphanage, this was the one which brought home to the children the unhappy realisation that they were in an institution and not in a caring home.

Her own unhappiness remained with her until she was able to put the picture of the orphanage from her mind and think once more of the reunion she would have with Gideon when they met again, in the Crimea.

III

For the nurses and sisters who had been on the whirlwind tour of Valetta, it was almost a relief to be back on board the *Tigress* and allowed to sink on to their uncomfortable bunks, hot and exhausted from their exertions.

Their relief was short-lived. The ship put to sea despite worsening gale force conditions and the unhappy passengers were soon to experience the worst weather the Mediterranean had known in living memory.

Gwen proved to be particularly prone to seasickness, but while she lay on her bunk, convinced she would die, Alice was quite happy to take care of Verity. She would make up stories to tell the young girl in order to help her forget the movement of the ship and the terrifying sounds of the storm and violent seas battering against the ship's hull.

At the height of the storm water began pouring into the hold and the hard-pressed crew battened down the hatch, leaving the women and children with no light and little air, and unable to leave the hold to obtain food.

The latter was a hardship to no one. Even had food been readily available there was not a single woman or child who could have faced the prospect of a meal.

When a brief lull in the severe weather occurred on the third day, the hatch was opened and Alice struggled to the deck with some of the hardier women to collect provisions for the unhappy passengers. She then accompanied Sister Frances when she went to the captain to demand that he come and see for himself conditions in the hold.

They were truly appalling. The weather had been so bad that the women occupying top bunks, fearful of being flung out in the terrifyingly rough weather, had slept on the floor of the hold. Unfortunately, despite the hatches being closed, water still found a way in and the floor was awash, soaking those who lay there. In addition, many women had been violently ill. This coupled with a lack of air made conditions in the hold unbearable.

Agreeing that such a situation was intolerable, the captain ordered that the women be transferred immediately to the larger and airier after hold. It was drier there and, away from the bows, the ship's motion would seem less violent in heavy seas.

Despite the change to more comfortable surroundings nothing could altogether negate the effects of the rough weather and by the time Constantinople was reached half of the *Tigress*'s passengers were incapacitated.

Alice was younger than any of the nurses and survived the ordeal better than most, but when she stepped ashore on the Constantinople dockside with Verity in her arms it seemed to her the land had taken on the movement of the sea. She needed to support herself against a nearby warehouse until the world about her ceased its nauseous gyrations.

When it was considered the unhappy passengers were able to walk without staggering too dramatically, they were taken to Scutari. Here they struggled to climb the hill to the hospital where the nurses were to work and the families would be housed until they were able to continue their journey to the Crimea.

The hospital was a huge building that had until recently been an army barracks and was one of the ugliest structures Alice had ever seen. When they went inside she discovered it was even worse than it appeared from the outside.

Built around a quadrangle which had served as a parade ground and was now piled high with a wide variety of rubbish, the wards and corridors were equally dirty, filth of all kinds being strewn everywhere.

The soldiers' families were taken to a wing given over to women and children, and when Alice handed Verity back to her mother both women promised not to lose touch with each other for as long as Gwen remained at Scutari. The nurses were split into groups, each allocated quarters in one or other of the towers situated at the four corners of the huge building.

There were no tables or chairs in the room where Alice and the Sellonites would live and they had only mattresses placed on the floor on which to sleep. There was strong evidence that the roof leaked whenever it rained and the women would soon learn they shared the room with colonies of rats, mice and fleas.

Alice and the sisters commenced cleaning up immediately and with a box serving as a table they utilised rolled-up mattresses as chairs and sat down to enjoy milkless tea and stale bread, sent up to them for their first meal on Turkish soil.

The nurses were not happy with their accommodation, but when they were shown round the hospital they realised they were much better off than the soldiers occupying the barrack-room wards. The place was already overcrowded, with not enough beds for the patients already there, yet within a few weeks there would be more than two thousand sick and wounded soldiers crowded into a building quite unsuitable for the use to which it had been put.

However, by the time the huge influx arrived the indomitable nurses had managed to clean up much of the interior.

Now Alice made a discovery that was to be a deep disappointment to her. Everyone she had spoken to before leaving England had been extremely vague about the proximity of the Crimea to Scutari, but she had been given to understand they were 'close'. It was only after she arrived in Turkey that she learned that the Crimea, where she had hoped to meet Gideon, was a three-day journey away, on the other side of the Black Sea.

She wondered whether she might be able to take passage there to find him, but, so far, she had met with no one she felt she could ask – and soon she was given something else to think about.

Wounded survivors from the glorious yet pointless 'charge of the Light Brigade' had reached the hospital more than a week before, but now casualties were arriving from an even bloodier battle.

The nurses had been told to expect two thousand wounded men, but that figure was soon overtaken by the huge numbers of diseased soldiers arriving at Scutari on almost every ship that docked from the Crimea.

Cholera was certainly present among the new arrivals. So too were scurvy and typhus. As winter storms increased in intensity, conditions in the Crimea deteriorated still further and the unfortunate soldiers besieging Sebastopol fell victim to the weather in ever-increasing numbers. Some began collapsing from the effects of severe exhaustion.

But it was not only combatants who were brought to the hospital. Wives and young children who were with the army in the Crimea were suffering too. It was these, a number of orphans among them, whom Alice found herself caring for. It was here too she was reunited with Gwen. The soldier's wife was also looking after children who had lost their parents.

These orphans would be sent back to England and cared for in military orphanages, unless relatives could be found to take responsibility for them. Both Alice and Gwen worked hard to try to find out whether the children knew of such relatives but it was not easy. Most of the children had spent the whole of their young lives 'following the drum', and had never met grandparents, aunts or uncles.

Many of the children were suffering from shock and malnutrition in addition to their other problems. Alice, Gwen and a qualified nurse were kept occupied tending to the needs of these unfortunates for all of the day and sometimes much of the night too.

One night, after cuddling one small boy until he sobbed himself to sleep, Alice sought out Gwen. As the two drank tea together in the semi-darkness of the hospital's orphans' ward, Alice was unable to contain the emotion she felt and blurted out her distress to the soldier's wife.

'It's the thing that worries me most, Alice,' Gwen confessed. 'What would happen to Verity if she were orphaned? She's not old enough to tell anyone about her grandparents and explain where they live.'

'Don't even think of it,' Alice said. 'The small boy I nursed to sleep was tragically unlucky. You and Verity are going to join your husband. You'll support each other and be all the stronger for it when you eventually get back to England.'

'I sincerely hope so, Alice – but I'm trying to be realistic,' Gwen persisted. 'No doubt the mother and father of that little boy never believed they would leave him an orphan. But with them gone, what will happen to him now? Even if he *does* have family in England, is there anyone who knows anything about them – and would anyone care anyway? No, it's far easier for all concerned to put him in an orphanage and forget about him – and let grieving relatives believe he died along with his ma and pa. It's understandable, I suppose, once you've seen the chaos here. No one knows the surnames of half the orphans.'

When Alice began to protest, Gwen asked sharply, 'Do you know the name of the boy you've just been caring for?'

Alice was forced to admit she knew him only as Harry.

'There you are,' Gwen said. 'That's exactly what I'm talking about. It frightens me to think of what could happen to Verity if Jack and I were both to die out here . . . no, Alice, don't tell me it won't happen. We all know it could.'

Alice could think of no argument to counter this and, after a few minutes of awkward silence between them, Gwen said, 'Before Verity and I leave Scutari I'm going to have a medallion put on a chain round her neck. It will say that if she's found, *you* are to be notified. If I do that, will you see that she's given into the care of Jack's ma and pa, in Cornwall, and not put into an orphanage . . . please?'

Alice was about to make some reassuring remark about nothing happening to Gwen and her husband, but she realised in time how deadly serious the other woman was.

'Of course I will, Gwen, but I hope it will never be necessary. Let's talk of more cheerful things. Have you heard yet when you'll be able to take passage to the Crimea and join Verity's pa . . . ?'

IV

Just when it seemed to Alice that the situation in Scutari could hardly be worse, a great storm swept the length of the Black Sea. The damage caused to the hospital was serious enough, but in the Crimea it was a disaster of unimaginable proportions. Ships containing stores, food and ammunition were dragged from their anchorages and sunk, while the tented hospitals close to the battlefront were obliterated. In addition, the countryside around Balaklava became an untraversable quagmire and the flooded trenches of the troops investing Sebastopol were cut off from the rest of the world for a while.

Cold, hungry, fevered and desperately short of supplies, the British soldiers were in no fit state to fight even the most minor of skirmishes. Had the Russians taken this opportunity to attack, the British army would have suffered a humiliating defeat, yet the ageing generals conducting the war seemed incapable of taking any decisive action. To those reporting the conditions, and to many junior officers, it was clear that if things did not rapidly improve there would no longer be a British army in the Crimea when fighting resumed in the spring.

The huge influx of invalids, together with their wives and children, threatened to overwhelm the resources of the two hospitals now in use at Scutari. It was not long before Alice and Gwen found they were working on their own among the families of the soldiers.

One day, a week after the great storm, two events happened to colour Alice's life. The first was the receipt of not one letter from Gideon but *three*, the earliest written while Gideon was still travelling to the Crimea. Alice realised they had both followed exactly the same route through France.

The letter told Alice of the places he was seeing, which she recognised. But, and far more important to her, he hoped she had given the question of marriage to him serious thought and would soon give him an answer.

Alice finished reading the letter with a warm glow inside her.

She had feared he might change his mind when he had been given time to think about his proposal to her.

The second and third letters followed the same theme, but they also described life in the Crimea, and the appalling conditions existing there.

Alice realised that had she not come to Scutari, his description of what was happening in the Crimea would have meant little to her. Now she knew only too well the conditions he had encountered. She was distressed to know he was sharing many of the experiences suffered by the soldiers and families she was looking after now.

She was still preoccupied with thoughts of Gideon when she went down to the quayside that afternoon to see five of the orphans she had been caring for board a ship bound for England – and an orphanage. Little Harry was among them. They were travelling with a number of widows and other children, as well as many seriously wounded soldiers, returning to England to be discharged from the army.

Alice had grown fond of the orphans in her care and she promised them she would try to find where they were and pay them a visit when she herself eventually returned home to England.

The orphans had embarked and the ship was preparing to cast off when a man approached Alice, who was waiting to wave goodbye to them.

Indicating her uniform, he said, 'Good day to you, miss. It does a man's heart good to see a pretty young woman so smartly attired. Are you one of Miss Nightingale's nurses from the hospital?'

'I work at the hospital and came to Scutari with the Sisters of Mercy,' Alice replied, wary of this well-dressed stranger, 'but I'm not a qualified nurse. I'm just seeing some orphaned children off on their voyage home. I've been taking care of them up at the hospital.'

'Then you're an angel indeed.'

The man's ingratiating yet overly bold manner put Alice on her guard, but she said nothing.

'No doubt your charges will be greatly relieved to escape the horrors of Scutari. Conditions must have come as quite a shock to the sisters too.'

'They *were* pretty horrible when we arrived,' Alice agreed, 'but things are very different for us now. Miss Nightingale quickly saw to that.'

'I am delighted to hear it,' said the man. 'I must come up there and see conditions for myself. Perhaps you could show me round?'

Aware of Alice's uncertainty about him, he smiled. 'I should have introduced myself right away. I am William Turnbull, war correspondent for a London newspaper. I arrived back from the Crimea only yesterday and came down to the docks today to arrange for some of my photographs to be taken to London. I must take your photograph sometime, miss. The readers of my newspapers would be delighted to learn that a young woman like yourself is here, helping the orphans of our unfortunate soldiers to forget the horrors of war.'

'I have a friend in the Crimea,' Alice said, wishing to change the subject and picking up on the fact that Turnbull had recently returned from the peninsula. 'He's helping to build a railway there.'

'I knew two railway builders while I was in the Crimea,' Turnbull said. 'Indeed, as far as I am aware, they were the only ones there. What's the name of your friend?'

'Gideon,' Alice replied, 'Gideon Davey.'

'Gideon!' repeated the correspondent, 'I know him well. Indeed, I travelled from Scutari to Balaklava with him when he first arrived from England.'

Despite her reservations about the correspondent, Alice was thrilled to meet someone who knew Gideon and had met him so recently. 'How is he? Is he well?'

'Had you been here an hour earlier you could have asked him yourself,' William Turnbull said. 'Unfortunately, he is no longer in the Crimea.' Pointing to a vessel that was fast disappearing westwards, he added, 'He's on that ship, returning to England.'

147

Observing Alice's dismay, Turnbull seized what he thought was a golden opportunity to advance his acquaintanceship with her. 'Don't be too disappointed, my dear. You must allow me to show you some of the sights of Constantinople and tell you of the sort of life your friend has been leading in the Crimea. I am sure you will find it of the greatest interest.'

12

I

Gideon reached London by train on a cold November evening when the city was held fast in the grip of a choking, sulphuric fog. It caught at his throat, irritated his eyes and, muffling traffic sounds, gave the capital city an air of unreality.

He and Robert Petrie had not met each other as had originally been planned. Petrie's injuries had been slow to heal, causing him to spend longer than intended in France. He had telegraphed for Gideon to return direct to England and recruit the men they would require to build the railway, hoping they might still meet up along the way.

However, by the time Gideon had completed his task in the Crimea, Petrie was sufficiently recovered to set off to see for himself the problems confronting the railway builders. His ship and the vessel on which Gideon was a passenger passed each other in the night somewhere between Constantinople and Malta.

Gideon was dismayed by his estimate of the cost of the railway. Together with the shipping bills that would be incurred it totalled an enormous sum of money. The amount of materials and equipment that would be necessary was in itself quite awesome, but it had been checked and agreed by Ranald MacAllen and when it was telegraphed to Petrie the contractor had accepted it without dispute.

Gideon's estimate included more than 2,000 tons of rails, 6,000 sleepers, shiploads of timber for bridges and viaducts, engines, cranes, rolling stock and horses, as well as clothing and supplies for the navvies and the tradesmen who would be working with them.

It would require a whole fleet of ships to carry the men and materials to the Crimea. Fortunately, other contractors had become involved in the ambitious project with Petrie, one of whom owned a shipping line. Even with this advantage it would be necessary to charter a number of additional ships. In order to increase the fleet still further, four steamships had been purchased especially for the venture.

Aware of the hazards likely to face the navvies, Gideon had also requested that a large number of revolvers be supplied for them, to be used if their efforts attracted the attention of the enemy. The suggestion that untrained and unruly navvies be armed had met with opposition from senior army officers when the request was received in London, but they had been over-ruled by government members.

Despite the London fog, Gideon set out that evening for a prearranged meeting with the other contractors and men closely associated with the Crimean project.

The discussions lasted for many hours, but by the time they came to an end everyone present was in full agreement on the plans for the Crimea railway and Gideon was complimented on the comprehensive report he had prepared whilst travelling back to England.

In the report he set out the difficulties posed by the terrain, the weather – and the Russian army. But he assured the backers that with two hundred and fifty navvies he would have the railway built in record time, whether or not the help promised by an exhausted army was forthcoming.

Before leaving the meeting he arranged for the contractors to advertise for experienced navvies interested in working on the proposed railway to apply for interviews to be held in London in a week's time. He would return to interview them personally – but first he was going to Cornwall.

Gideon travelled to Plymouth by train the next day and spent an impatient night at a Plymouth inn. Early the next morning he boarded a coach that reached Liskeard, the present railhead for the Cornish railway, a few hours later.

Hiring a horse from one of the town's stables, he set off to where his gang was working and, after being subjected to a rowdy welcome, was impressed to discover they had completed the task he had set for them, well ahead of time, and were currently working on another contract which was also nearing completion.

The men were delighted to see him again – Sailor particularly so. Gathering his gang about him, Gideon told them where he had been and what was planned there, and pointed out the dangers and exceptional difficulties that would be encountered by any navvies building a railway in the Crimea. Despite this, when he called for volunteers to accompany him on his return to the peninsula, every man of his gang asked to be placed on the payroll.

Walking away from the meeting with Sailor, Gideon said, 'I knew I could rely on the men – and on you too, Sailor. You've done well in my absence. Now I want to go and see Alice before I do anything else. How did she react when you told her I was going to be away for a while?'

Suddenly ill at ease, Sailor replied, 'Getting your message to her proved more difficult than I thought, Gideon. When I arrived at the rectory the first time, I caught the lad who'd daubed paint on the door . . .'

He described what had happened on his first, abortive visit to Alice, and told Gideon of the demolition of Billy Stanbeare's cottage by the navvies. Gideon was delighted that Sailor had caught the young village boy red-handed and approved of the way he had dealt with Billy Stanbeare.

'I'm proud of you, Sailor. You and the men will have taught

Stanbeare that I meant what I said about making certain he was punished for any action he took against Alice – whether or not I was still in the area. What happened when he discovered what you'd done?'

'He and his father swore out a warrant for your arrest, so it was just as well you weren't around. I believe they both got a tongue-lashing from the magistrate when he learned you were out of the country at the time. He said that as far as he was concerned, the collapse of the cottage must have been caused by bad maintenance and wet weather!'

Gideon chuckled. 'Well done, Sailor. But what about Alice? Did she seem upset when you told her I had gone away?'

Observing Sailor's discomfiture, Gideon demanded, 'You *did* tell her?'

'Yes, of course I did. The only thing is . . . Well, I couldn't tell her that night, and it was another week before I was able to get back to Treleggan. By the time I called in to speak to her a new parson had moved into the rectory. His wife came into the kitchen when I was talking to Alice and raised hell.'

Alarmed, Gideon asked, 'Haven't you been back there since, to check that Alice is all right?'

'In view of what the parson's wife had to say about navvies, I thought it best not to. Besides, we've been working a bit farther away from Treleggan than we were then.'

Gideon realised that life must have changed a great deal for Alice during the time he had been out of the country. Thankful that he had chosen to hire a horse, he decided to ride to the rectory right away and learn what was happening. After promising Sailor he would spend the whole of the next day with his gang and explain more of what would be required of them in the Crimea, he set off for Treleggan.

Along the way, the more he thought of what Sailor had told him of his last visit to the rectory, the more anxious he became. Navvies had a somewhat unsavoury reputation – and Gideon had to admit it was not entirely without foundation. A rector's wife was not likely to take kindly to a housemaid who entertained them in her home.

In spite of such concerns, Gideon was excited at the prospect of being with Alice again. They would still need time to get to know a little more about each other, of course, but he did not doubt they could be very happy together. If she agreed to marry him, he would be able to solve any problems she had right now.

The fact that he would need to return to the Crimea would certainly make things a little more difficult, but he had been able to save money as a result of his work as a ganger and Robert Petrie had given him an advance on the very generous salary he would be earning in the Crimea. Alice would not need to continue working at the rectory if the new parson's wife was creating problems for her.

When Gideon reached Treleggan, he attracted considerable attention from the villagers. It was hardly surprising. Few strangers on horseback found their way to the remote village and many of those who saw him recognised him as the ganger who had taken an interest in Alice. They wondered what he was doing back in the village riding a horse and dressed in the clothes he had worn for his travels in Europe. Clothes more befitting a gentleman than a navvy.

In view of Sailor's experience with the new rector's wife, Gideon had decided he would not go to the kitchen door and risk compromising Alice once more. Instead, he tugged the bell-pull at the front door of the rectory, intending to ask permission to speak to Alice.

It was possible, of course, that Alice herself would open the door to him . . . and he found that an exciting thought!

Being unaware of the Reverend Bushell's background, he was taken by surprise when the door was opened, not by Alice, but by an Indian maid. She seemed equally surprised when he said he had called to see Alice.

'Who is it, Shabnam?' The ringing question came from the direction of the kitchen.

'It is a gentleman. He is asking for Alice.'

'Oh, is he!'

There was the sound of hurrying footsteps, and, brushing

aside the Indian maid, Harriet confronted Gideon.

She was taken aback at his appearance and he correctly guessed she had been expecting to see someone dressed in the work-clothes of a navvy. Taking advantage of her surprise, Gideon called upon the mode of speech taught to him by his mother. It was something he had allowed to lapse when working with his gang.

'I apologise for troubling you, ma'am, but I wonder if I might have a few words with Alice, your housemaid?'

Recovering some of her composure, Harriet replied, 'I am afraid that is not possible.'

Believing she was merely being difficult, Gideon said, 'I will not keep her from her duties for more than a few minutes, ma'am . . .'

Before he could complete the sentence, Harriet said in a clipped voice, 'Alice has no duties here. She left my employment many weeks ago. Left with no thought of the inconvenience it would cause to my husband and me, as complete strangers to the area.'

Dismayed, Gideon said, 'Alice has left? Do you know where she has gone? Where I can find her?'

'No – and what is more, I do not care. It is all very well for the rural dean to come here telling me what a loyal servant she was to the late Reverend Markham. She showed very little loyalty to me. Had it not been for Mister Stanbeare, I do not know how we would have managed.'

'Would that be Mister Stanbeare senior?'

Gideon was wondering where he could start looking for Alice. He was aware that Henry Stanbeare had certainly never been kindly disposed towards her, but he might have an idea where she had gone.

'No, his son, Billy.' Harriet broke in upon his thoughts. 'Is he a friend of yours?'

'I wouldn't say that, ma'am, but we do know each other.'

Despite his concern for Alice, Gideon could not help wondering what motive Billy had for helping the new rector and his wife – and Gideon was convinced he would have one.

'Well, if there is nothing else, I have matters to attend to,' Harriet said frostily.

'Yes, of course. Thank you.'

Gideon turned away, then swung back abruptly as a sudden thought came to him. 'I sent a number of letters to Alice. Do you have them?'

'No, the rural dean said I was to forward them to him. I did so.'

'Where will I find the rural dean?' Her words had rekindled Gideon's hopes. If the rural dean had asked for Alice's letters to be sent to him, he must know where she was. She might even be working in his household.

'You will find Reverend Brimble in Bodmin – but do not ask me to direct you to his house. Reverend Bushell and I have never been invited there.'

With this resentful remark, the door was closed, leaving Gideon with an impression that he had just been talking to an unhappy and embittered woman.

II

Taking a direct route to Bodmin, Gideon urged his horse along a faint path that led across Bodmin Moor. His way took him close to Tabitha Hodge's kiddleywink where the old widow was in her vegetable garden, using a long-handled spade to dig up potatoes which she then threw into an ancient wooden bucket.

Seeing Gideon approaching, she straightened up, one hand held in the small of her back, the other clutching the spade for support.

'Hello, Widow Hodge.' Gideon pulled his horse to a halt and greeted her.

'Well, if it isn't Alice's navvy . . . or him as used to be a navvy. Unless I'm very much mistaken he's gone up in the world and is a gentleman now.'

Gideon grinned. 'I'm not quite a gentleman yet, Widow

Hodge . . . but give me a few more years and I'll have come close to it.'

'Then you'll no doubt be far too grand for the likes of young servant girls.'

Suddenly serious, Gideon said, 'Not if you're talking of Alice. I've just been to the rectory at Treleggan to find her, but there's a new parson there. His wife told me Alice had walked out on her.'

'I think you'd better get down and come into the house to hear the true story. There's a rope tied to the tree over there. Tether your horse to it. There's still enough grass around to keep the animal content for a while. You can bring this bucket of potatoes in for me when you come.'

'I can't stop for long, Widow Hodge. I want to get to Bodmin and find a Parson Brimble. It seems he knows where Alice is. If she's not somewhere nearby I'd like to be back to Liskeard before it's dark.'

Hobbling ahead of him, Tabitha said, 'You'll be in Bodmin a good half an hour before dark – but you won't find Alice there. I believe Parson Brimble found work for her outside Cornwall. I'm glad you're still seeking her though, even if you are halfway to being a gentleman now. She's a good girl, that one. Too good to be employed in a kiddleywink, as I told her when she came here begging me to give her work.'

'Alice wanted to work here?' Gideon looked at Tabitha in disbelief.

'I didn't say she *wanted* to work here. The girl was desperate and didn't want to be too far from Treleggan in case you came back looking for her. But I put a good meal inside her and sent her on her way to seek help from Parson Brimble . . . and she didn't "walk out" as you were told by that stuck-up madam they've got in Treleggan rectory now. Alice was told to go when the rector's wife had hardly set foot in the place. From what I hear it was Henry Stanbeare who set her and the parson against Alice before they'd even met her.'

As she spoke, Tabitha was kicking off the heavy boots she had been wearing in the garden. The task completed, she slipped on a pair of soft, comfortable shoes and continued, 'Henry Stanbeare might be looked up to as an important churchman by them as don't know him well, but if he ever gets beyond the gates of heaven then there's no more justice in the next world than there is in this one. But, here, get this drink inside you. There's a cold wind blowing across the moor today – and this is from my "special" barrel. You won't find brandy like it anywhere else in Cornwall – not any more, you won't.'

The drink was certainly warming and left a comforting glow on his palate. After agreeing with Widow Hodge that it was top quality brandy, he asked, 'Are you quite certain it was *Henry* Stanbeare who turned the new parson's wife against Alice? When she spoke to me about Alice letting her down, she said it was Billy who'd been so helpful to her. I must admit it surprised me. I've never thought of him as the type of man to go round offering help to people – unless he stood to gain something from it.'

Tabitha chuckled. 'Young man, you're a better judge of character than her at the rectory, as she'll learn to her cost. But, as is so often the case, it won't be the one who deserves it most who'll pay the highest price.'

Puzzled, Gideon asked the old widow what she meant.

'The ones who'll suffer for her trust will be those two Indian girls she's got working in the rectory, you mark my words.'

'You mean . . . Billy's bedding one of them?' Gideon queried.

'Not one, but both,' was the reply. 'From what I hear, he takes them off, one at a time, to show them where they can buy milk, meat, vegetables, and so on – but takes 'em by a roundabout route. A Cornish girl would see through him right away, but I doubt if these foreign girls have ever come across anyone quite like Billy Stanbeare. Mind you, I'm only repeating what I've heard said in here – but I've no cause to doubt it. Some of the miners even say they've seen Billy at it.'

'I don't doubt that it's true,' Gideon agreed. 'But I must get on my way now, Widow Hodge. Thank you for the brandy, it's every bit as good as you said it was. I'll come back for another one of these days.'

'What happens if Parson Brimble tells you where you can find Alice and you meet up with her again, what will you do then?'

'The same as I would have liked to have done before I went away. Persuade her to marry me.'

Looking shrewdly at Gideon, Tabitha said, 'I wouldn't let Alice work here when she asked me. Is she going to be any better off living in a camp with the sort of women who follow navvies around the country?'

'Alice will never need to live in a navvies' camp,' Gideon said firmly. 'If things go well when I return from the Crimea in about six months' time, I'll be sub-contracting on a railway line somewhere and able to buy a house, possibly in Cornwall. Alice can either stay there, or come with me, as she chooses. If she comes with me when I'm working I'll be able to afford to rent a house for her, close to the line. That's assuming she accepts me, of course. She didn't when I first asked her.'

'Of course she didn't,' Tabitha said scornfully. 'My William had to ask me at least four times before I said yes, and then it wasn't until he'd agreed that I'd be the one to manage the money he brought home every month. I always made certain he had enough in his pocket to enjoy a drink or two at the end of the week, but not enough to get stupidly drunk, as too many miners did – and do. William never regretted the arrangement and we enjoyed a good life together.'

'I'll remember that when I bring home my pay, Widow Hodge.' Gideon smiled. 'Again, that's if she decides she'll marry me, of course!'

'She will,' Tabitha said confidently. 'When she called in here she was far more worried about you not being able to find her again than she was at not working for the new parson at Treleggan.'

'Thank you, Widow Hodge. When the time comes I'll make certain you have the first invitation we send out for the wedding.'

'Then you'll need to get married on the moor,' Tabitha declared firmly. 'I haven't left Bodmin Moor in all my ninety years. I don't intend stepping off it now – not even for Alice's wedding.'

When he reached Bodmin, Gideon made enquiries and soon found the Reverend Brimble's house. His tug on the bell-pull was answered by a maid, who called her mistress to the door.

Introducing himself to the rural dean's wife, Gideon explained that he was enquiring after Alice. Beatrice Brimble maintained a certain reserve in her manner towards him until he asked whether the letters he had sent to Alice had been forwarded to her.

'Of course!' Beatrice exclaimed. 'You will be the young man Alice was so concerned about when she left Treleggan. Did you not ask her to marry you?'

'I did,' Gideon confirmed. 'She didn't accept then, but I hope to be able to persuade her to change her mind when we meet up again.'

'Please come in,' Beatrice said. 'I will call my husband. He is in his study working on church accounts. He will be happy to forget them for a few minutes and he knows more of what is happening with Alice than I.'

Inside the house, Gideon received an enthusiastic welcome from Digger, and as he was making a fuss of the small dog he looked up to see Parson Brimble coming along the passage towards him.

Extending a hand to Gideon, Harold Brimble said, 'Digger seems to remember you, dear boy – and I am very pleased to meet you, too. I am happy to know Alice has found someone to take care of her. The poor child has not had the happiest of lives. Marriage will change everything for her, I have no doubt.'

'I hope Alice feels the same way and agrees to marry me.'

Harold Brimble looked momentarily puzzled before saying, 'Young women are sometimes reluctant to commit themselves to matrimony until they feel they know as much as possible about their future husband – but Alice seemed quite certain of her feelings when she was here. However, Beatrice tells me you have a query about the letters you sent to her. There were three, as I remember. They were forwarded to me by the Reverend Bushell at Treleggan. I passed them on to the Reverend Mother at St Dunstan's Abbey in Plymouth and I know she is meticulous in such matters. Has Alice not received them?'

Now it was Gideon's turn to be confused. 'I don't know, sir. I haven't seen Alice since a few weeks after Parson Markham died. I don't even know where she is. You see, I've been out of the country, helping to survey a railway we intend building in the Crimea.'

Harold Brimble looked at Gideon in stunned disbelief. 'You mean . . . you didn't know? You haven't met Alice? My dear boy, what a dreadful confusion there has been. You see . . . Alice left for the Crimea with a large party of nurses and Anglican nuns, expecting to meet up with you soon after they arrived. She and the others are working in a hospital at Scutari. You must have passed almost within sight of one another. And now you are separated by thousands of miles! How very, very distressing – for both of you!'

13

I

'Have you heard the news? Another forty-seven nurses have arrived on a ship at Scutari this morning.'

The surprising information was given to Alice by Gwen, who had not yet been able to arrange a passage to the Crimea in order to join her husband.

'We can certainly do with extra help,' Alice replied. 'If it was a hundred extra it wouldn't be too many.'

'That isn't what Miss Nightingale said.' Lowering her voice in a conspiratorial manner Gwen added, 'One of the orderlies was nearby when she was told the news. He said she was *furious*! She wanted them all sent back to England immediately.'

'Why on earth should she even suggest such a thing?' Alice said scornfully. 'Everyone knows how difficult it is to cope with the sick and wounded who are coming in all the time.'

'She told the chief medical officer that unless *she* says who can or can't work here we'll start getting the wrong sort of women in the hospital again and things will go back to the way they were before she was put in charge. She said all her nurses are hand-picked and she intends to keep it that way.'

'Well, Miss Nightingale didn't hand-pick me, or you, and

we're working as hard as anyone else. Besides, the state most of the soldiers are in when they arrive here, a cheerful smile from a woman – any woman – is just as important to them as having a trained nurse dress their wounds, or hand out their medicine.'

'That's what I said to the orderly,' Gwen said. 'It seems the chief medical officer said much the same thing to Miss Nightingale. Although she's certainly not happy about it, she's agreed to keep the nurses here, "for a trial period".'

'I should think so!' Alice declared. 'We've eight nurses sick as it is – and three more have been sent home. That's the reason you and I are working on the wards. Nobody's told us we can't do it just because we haven't been "hand-picked".'

What Alice said was the truth. She and Gwen were now working on a soldiers' ward, where the need was greatest, doing as much as the trained nurses for their patients: making certain they took their medicine and ensuring they were as comfortable as it was possible to be in the crowded and trying conditions. They would also occasionally change the dressings of wounded soldiers if the surgeons were occupied elsewhere, and generally tried to bring a little cheer into the lives of the patients in their care.

But with the arrival of the additional nurses, things quickly became worse for Alice and Gwen. In a reallocation of duties, the Honourable Josephine Wotton, the nurse who had made things difficult for Alice on the voyage from England, was put in charge of the ward where the two young women were working.

Nurse Wotton was horrified that two 'unqualified' young helpers were carrying out nursing procedures. Twenty-four hours after she assumed her duties, she called them both to the small room which served as the ward office. They were informed they must no longer act as nurses, but would be employed instead as ward orderlies, engaged mainly on such menial tasks as cleaning floors and bed-making.

'But that's a total waste of our skills,' Alice protested. 'Recuperating soldiers can carry out orderly work. Gwen and I have been taking care of the men for weeks now without a complaint from the surgeons, or from the men themselves – indeed, we've had nothing but praise from everyone.'

'I do not doubt you have performed your tasks to the best of your abilities,' said Nurse Wotton stiffly. 'However, it is work for which you are not qualified. The matron who is now in charge of this wing of the hospital agrees with me. You will cease all nursing duties immediately.'

Alice was aware that the newly appointed matron was a friend of Nurse Wotton and would have been influenced by the baron's daughter. Nevertheless, she did not accept the decision without argument. She pointed out that she had acquired considerable experience of working with cholera victims before coming to Scutari and, after being given training by Sister Frances, had gained first-hand experience of treating injured men. She was probably more highly qualified in all aspects of nursing than any of the newly arrived contingent.

Her arguments were futile. Nurse Wotton had decided that Alice had been encouraged to 'put on airs' by taking care of patients. She intended ensuring the ex-housemaid should return to duties more in keeping with her station in life.

Gwen did not bother to argue. She told Nurse Wotton she had dealt with more wounded soldiers than the Honourable Josephine was ever likely to meet and warned her that if she dispensed with the skills of Alice and herself, many good men were likely to die. Furthermore, Gwen had no intention of remaining to see it happen. She would move back into the wing occupied by soldiers' families, where she would be able to spend more time with her daughter, and do nothing while waiting for the army to arrange a passage to the Crimea for them both.

That evening, after completing her orderly duties, Alice sought out Sister Frances and poured out her troubles to the down-to-earth nun.

Sister Frances was sympathetic, but admitted there was nothing she could do about the situation. The order about not employing unqualified women on nursing duties had been made by Florence Nightingale herself before she was aware of the situation at Scutari. She was not likely to change it now there had been an influx of new nurses, no matter how efficiently Alice and Gwen had performed their duties.

What was more, Sister Frances could not even return Alice to her original business of caring for the needs of the Sisters of Mercy. Their numbers had dwindled through sickness and those remaining had been dispersed between the two hospitals operating in Scutari. Having an English maid could not be justified, especially as local Turkish women were now employed to perform menial tasks for the nurses.

Aware of the situation in respect of Gideon, Sister Frances pointed out that if Alice wished to remain at Scutari until Gideon returned to the Crimea, she had no alternative but to accept that she would need to work as a ward orderly.

Alice did not find the transition from nurse to orderly an easy one to make whilst still working in the same ward – especially when she watched the manner in which Nurse Wotton carried out her duties. The aristocratic woman might be 'qualified' but Alice thought her totally lacking in compassion. As a result, she twice found herself in trouble for easing the dressings of enfeebled soldiers when the nurse failed to heed their cries for help.

On the second occasion, Alice was taken before the matron and warned about her actions. If she stepped out of line again she would be dismissed.

Alice left the ward late that evening thoroughly unhappy. She had felt she was carrying out a worthwhile task when she was nursing wounded soldiers and found it difficult not to step in now if help was needed. It would be well-nigh impossible not to fall foul of Nurse Wotton yet again.

II

That night, as Alice was preparing for bed in a tiny room that had once been a store cupboard, there came a gentle knock at the door. Uncertain who might be calling on her at this time of night, Alice put her head close to the door and called, 'Who's there?'

To her great surprise, a voice she immediately recognised said, 'It's me, Alice. Gwen.'

Drawing back the bolt she had insisted be attached to the inside of the door to give her some security while she slept, Alice threw open the door and cried, 'Gwen, what are you doing here? There's nothing wrong with Verity?'

'No, she's asleep, in the family wing. One of the other mothers is keeping an eye on her,' Gwen said. 'But when I went down to the docks today to see if I could arrange our passage to the Crimea myself, two ships came in together, both loaded with hundreds of seriously ill and disabled men. There were so many it was obvious they would be left lying or sitting on the dock-side in the cold drizzle for hours. It was pitiful. Then I saw a woman going among them and handing out things that were being carried for her by two Turkish helpers. Tobacco, medicine, all sorts. A lot of the soldiers seemed to recognise her and were calling out for her to come and help them.'

'Who was she?' Alice could think of none of the nurses with whom she had travelled to Scutari who would leave the hospital and go to the dockside when sick and wounded men were being offloaded from the ships.

'She's a soldier's widow named Nell Harrup,' Gwen replied. 'In fact, she's been married and widowed three times. A soldier was her husband each time, and she's nursed them in China, India and the West Indies – that's why so many men recognised her down at the docks. The soldiers I spoke to said she's got a reputation for healing that's the envy of every doctor who's ever met her.'

'Is she one of the new nurses?' Alice asked.

'She would have liked to be, but Miss Nightingale wouldn't take her on – for the same reasons we were stopped from nursing, although according to Nell it's because she doesn't have a posh voice. She's a cockney, and speaks like one.'

'Then what's she doing out here?' Alice was puzzled. She had thought that apart from the new arrivals, the only nurses at Scutari were those who had been approved by Florence Nightingale.

'Nell had a small shop in London,' Gwen replied, 'selling all sorts of things, including some of the medicines she'd learned about on her travels. One day one of the officers she'd helped to get well in the West Indies – a titled gentleman – came into her shop. He'd been sent home from the Crimea and invalided out of the army. When he told her how bad things were here she said she'd like to come and help but, as I said, she wasn't accepted as a nurse. Anyway, one day this gentleman went off to meet some of his friends who were on their way back from the Crimea and soldiers from his old regiment were on the same ship. He was shocked by the state of them, and even more so when they told him the regiment had lost half its strength – most from sickness. He came straight back to Nell and said he'd pay for her to come out to the Crimea with her medicines and treat the men. But as well as being a good doctoress, Nell's a businesswoman. She suggested that if he put up the money to buy stock and equipment for an eating-house where she could also sell her medicines, they could become business partners. He agreed . . . and now she's here, with a ship anchored off Scutari full of things for her eating-house. Not only that, but after I'd spent a couple of hours helping her, down at the docks, she offered to take me to the Crimea if I'd work for her there. I told her I couldn't do that because once I got there I'd have a husband to look after – but I said that if she'd give Verity and me a passage on her ship I'd find a first class nurse to work with her.'

'Is this "first class nurse" . . . me?' Alice was not quite certain whether she was appalled or pleased.

'Of course.' Gwen hesitated a few moments before adding, 'I told her all about you, Alice, and how you were probably a better nurse than I was, but were being made to work as a ward orderly. She said she'd like to meet you.'

Alice was filled with uncertainty. She was not happy at being an orderly, but by working with the nurses recruited by Florence Nightingale she at least had semi-official recognition. If she tried

166

very hard to do as she was told she would be able to remain at Scutari until Gideon returned to the Crimea.

On the other hand, she had already received two warnings and was likely to be dismissed for no reason at all if she did anything to upset Nurse Wotton – and Nell Harrup was going to the Crimea, to where Gideon was due to return.

'At least come and speak to Nell, Alice. I mean, you might not like her – although I'm pretty certain you will.'

'Where is she now?' Alice asked.

'I left her helping the soldiers down at the docks. Another ship was coming in just as I was leaving, so she'll be there until late. But if you want to speak to her don't leave it too long. She's only waiting for the ship loaded with her things to be given permission to berth in Balaklava, then she'll be leaving right away. She told me it's on the peninsula where nurses can save the most lives, not here.'

'The trouble is, I'm working tomorrow,' Alice said, '. . . every other day too, if Nurse Wotton has her way.'

'Then come with me to the docks now,' Gwen said. 'Verity will be all right with the woman who's looking after her – she has a little girl of her own. Come on, Alice, before it's too late. Who knows, she and I might be gone by tomorrow.'

Alice was expected to be at work early in the morning, but she realised that, as Gwen had pointed out, if she put off a meeting with Nell Harrup it might be too late.

Making up her mind, she said, 'All right, Gwen, let's go and find her right now.'

III

It was the first time since arriving in Scutari that Alice had left the hospital after dark. At first she felt nervous as she and Gwen walked through the narrow, shadowy streets. All about them was the smell of cooking fires, food, and the heady aroma from hookahs, smoked by men seated on the steps of the small houses on either side of the streets. However, the docks were reached without incident.

Nell was still on the dockside. So, too, were hundreds of ill and exhausted soldiers, who had arrived on the latest ship to berth at Scutari.

It was an eerie scene, lit only by the flickering yellow glow from oil-burning lanterns, yet even in this light Alice could see that many of the men showed the symptoms of cholera. It alarmed her. The hospital was already packed with far more cholera cases than it could cope with.

Gwen was scanning the dockside in a bid to find the woman she wanted Alice to meet and suddenly she exclaimed, 'There she is . . . there's Nell.' She pointed to where a woman kneeled beside the stretcher on which a dying man had been carried ashore.

Nell Harrup was not at all as Alice had imagined she would be. Big both in height and weight, she was a woman in her late fifties, wearing a bright yellow dress. On her head was a poke bonnet decorated with artificial flowers and tied with ribbons that matched the dress.

As Gwen hurried towards her, followed more slowly by Alice, Nell Harrup drew the blanket covering the soldier's body up over his face before wearily rising to her feet.

Shaking her head, she turned towards the two young women and Alice saw tears glistening in her eyes. Recognising Gwen, Nell said, 'Poor lad, he died thinking I was his mother, but I suppose it gave him some comfort in his last minutes. I hope so. He was hardly more than a boy.'

Pausing for only a moment in deference to the other woman's feelings, Gwen said, 'Nell, this is Alice, the girl I was telling you about. The one who should really be nursing instead of cleaning floors.'

Putting thoughts of the dead soldier behind her, Nell peered closely at Alice before asking, 'How much do you know about nursing, love?'

'Enough to recognise that many of the men here are suffering from cholera,' Alice replied.

'Oh? And where did you learn about cholera?'

'In a Plymouth orphanage. I was helping Sister Frances of the Sisters of Mercy nurse children during an outbreak. Later we nursed victims in the town, when cholera was discovered there.'

'I know of the Sisters of Mercy. They had a place in London, not too far away from my shop. They were well-meaning and helped the poor as much as they could, but they prayed too much for my liking. Do you pray over your patients, or do you get on and try to cure 'em?'

Momentarily taken aback by the woman's bluntness, Alice recovered sufficiently to say, 'I like to think of myself as a Christian, and was maid to a parson from the time I was a young girl until he died, but I keep my prayers between myself and the one I'm praying to.'

Nell nodded, in apparent approval. 'Tell me, what treatment did you give to those who had cholera?'

Alice began to list the medicines that Sister Frances had taught her to use, but Nell cut her short.

'The mustard and some of the other things are fine; most of the rest are as useless as a flannel belt. Someone suffering from cholera is usually calling for a drink for much of the time. I've found giving them water that's been boiled with cinnamon – and making certain they drink nothing else – does more good than all the other potions put together. Mind you, I do give them *some* of the medicines the doctors swear help, but I believe that if I gave them nothing but boiled cinnamon water they'd stand just as much chance of recovering. Talking of which . . . there's cinnamon water in the bucket standing over there, by the flagpole. One of the Turkish boys working for me is off boiling up some more. While you're here, let's see if we can make these poor soldier boys more comfortable . . .'

Alice and Gwen began moving among the soldiers, most of whom were desperately thirsty. The two young women distributed cinnamon water and did what they could to make them comfortable as medical orderlies worked steadily to move men from the dockside to the hospital.

It was more than two hours before the last of the sick soldiers were finally moved away. Not all made it to the hospital. Some twenty men with the stillness of death upon them were left behind, lying in anonymity beneath army issue blankets.

'You did well, girls, and the men were grateful to you. I heard more than one refer to you as "bloomin' angels". That's high praise from a soldier.' Nell gave the two young women a tired smile.

'I only wish we could have done more for them,' Alice said with feeling.

'You can,' Nell said. 'I'm told my ship will be allowed to sail for Balaklava at noon tomorrow. It's in the Crimea that there's the greatest need for women like us, with the skill and gumption to nurse the men back to health. Come with me. You'll do far more good there than here in Scutari.'

Gwen glanced questioningly at Alice, who looked undecided.

Thinking Alice's uncertainty might have something to do with money, Nell said, 'Our main task will be to care for the sick and wounded, wherever we find them, but I'll expect you to help me in my eating-house when we have quieter moments. Whatever it is you do there I'll pay you a good wage. Mind you, you'll earn it. You'll also earn the gratitude of the men you're caring for – and their families back home in England. Although that pays nothing extra you'll keep it with you for far longer than you'll keep the money, take it from me. Well, what do you say? Will you come with me?'

'How about it, Alice?' Gwen asked. 'Will you come to the Crimea and help Nell?'

While tending the men on the dockside, Alice had been thinking of what was being asked of her. She realised that what Nell said was true. If she went to the Crimea her work would be far more worthwhile than if she remained as an orderly in the wards of the Scutari hospital.

But it was the thought of Gideon's return which tipped the balance for her.

'Yes, I'll come with you to the Crimea, Nell. I'll speak to Sister Frances first thing in the morning and have my things

170

down here in time to catch the boat with you.'

Excited at the thought of what the next day would bring, Alice and Gwen were leaving the dockside when they were startled by a figure who moved out to meet them from the shadows beside the stables which housed the dockyard horses.

'A very good evening to you, ladies – or should I say angels? I have been watching you for a long time while you tended our poor soldiers. I am impressed – *very* impressed – by your compassion, patience and skills. I promise you that by the end of this week, through the pages of my newspaper, your names will be known to everyone in England who is following the inglorious progress of this calamitous war.'

It was William Turnbull. His words were accompanied by the distinctive aroma of strong drink on his breath.

Falling in beside the two women, Turnbull went on, 'May I walk along with you? We know each other, of course, but one likes to give our readers as much background knowledge as is possible of people in the news. Better still, why do we not all go to the Hotel Malvern? It has been recently opened by an Englishman and has wonderful views over the Bosphorus from the terrace. You can enjoy a well-deserved drink – something you will not get at the hospital.'

'I'm sorry,' Gwen apologised, 'I've left my little girl with someone at the hospital and I've already been away for far longer than I expected to be.'

'And I need to be up early in the morning to see Sister Frances and pack my things,' declared Alice.

In truth, she would have found some other excuse for not accepting Turnbull's hospitality had she not had a valid reason. Since their very first meeting, the correspondent had never attempted to hide the attraction he felt for her. Alice neither liked nor trusted him.

'That's a very great pity,' said the disappointed correspondent. 'Never mind, we can talk as I accompany you back to the hospital – and I will accept no argument about *that*. It is not safe for two young Englishwomen to be walking the streets of Scutari at night without an escort. Now, first of all, tell me what first brought you to Scutari; your future plans – and how you came to meet with Nell Harrup . . .'

'Are you quite certain this is what you really want to do, Alice? Have you thought about it very carefully?'

'Yes, Sister Frances. I believe there is much more I could be doing for wounded and sick soldiers, but I am not allowed to do it here.'

Sister Frances looked pained, but she nodded her head with a gesture of resignation. 'I agree with you, Alice. Unfortunately, as I have said before, it is out of my hands.'

'I am not blaming you, Sister Frances. You and I have worked well together in the past. I am sure we could do the same here if it were permitted, but it isn't. That's why I have made up my mind to go to the Crimea with Mrs Harrup.'

'Mrs Harrup has an impressive record when it comes to tending those who have need of help, Alice, but I regret that her morals are, shall we say . . . somewhat suspect? I believe she has had at least three soldier husbands?'

'I know nothing about her morals,' Alice said, 'but I've watched her treating wounded men on the dockside. She not only does it very efficiently, but I believe she really cares for them.'

'Very well, Alice, I trust your judgement. You go to the Crimea with my blessing – but I expect you to maintain your own high standards when you get there.'

Alice's confidence in the wisdom of her decision to go to the Crimea with Nell Harrup suffered a setback the following day when she learned that William Turnbull would be travelling with them. Gwen and Alice had told him of their plans the previous evening and he had wasted no time in asking Nell to allow him to take passage with them to the Crimea.

Observing the close attention Turnbull was paying to Alice, Nell took the first opportunity she could find to speak to her. 'I am not a prude, Alice, and I've lived too full a life to dictate to others what they should or shouldn't do. When we reach the

Crimea, what you do in your spare time is entirely your own business, but I want to make one thing absolutely clear. While I am paying your wages, your first duty will be to injured and sick soldiers. I expect you to be available at any hour of the day or night to go wherever you're needed, even if it means closing the eating-house in order to do it. That's far more important to me than making a profit. Do I make myself clear?'

Alice nodded her understanding, but later asked Gwen what *she* thought Nell meant.

'She meant exactly what she said, Alice. As long as you do what she's paying you to do, she won't question how you spend the rest of your time.'

'I think that's what was worrying Sister Frances when I told her what I intended doing. She was concerned that Nell's morals aren't all they ought to be. She is afraid I might fall into her ways.'

'You needn't worry about Nell's morals, Alice, but she's been around for long enough to accept people for what they are. She'll be paying you good money to nurse needy soldiers, and help her run an eating-house. Do that well and she won't try to run your life for you.'

When Alice expressed her concern that William Turnbull should be travelling with them, Gwen said reassuringly, 'Don't worry about him, he's all talk. If you were to take him up on half of what he suggests he'd be frightened nigh to death. Besides, you and I are sharing a cabin – and, for all her apparent broadmindedness, Nell won't let Turnbull step out of line.'

IV

Alice's fears about having to fend off Turnbull's attentions while they were on board the *Princess* proved groundless. She discovered, as had Gideon before her, that the correspondent was not a good sailor. The *Princess* had an uncomfortable wallowing movement when under way and Turnbull did not emerge from his cabin until the ship dropped anchor in the calm waters outside Balaklava harbour. The harbour itself was ridiculously tiny. Far too small for the purpose for which it was currently being used by the occupying forces.

Joining Alice, Gwen and Nell as the three women were gazing at the long, narrow inlet that formed the entrance to the harbour, a somewhat wan Turnbull asked, 'Have you ladies been able to make arrangements for accommodation in Balaklava?'

'Not yet,' replied Nell, 'but until I've spoken to senior army officers I won't know whether I'll be setting up my store in Balaklava, or closer to the siege lines.'

Turnbull looked at her in disbelief. 'But . . . it will *have* to be here, in Balaklava. You can't even consider going anywhere else. It would be far too dangerous!'

'Tell me, Mister Turnbull, where are the soldiers in most need of medical help and any comforts that can be provided for them?'

Taken aback by Nell's question, Turnbull replied, 'Why, up by the siege lines around Sebastopol, I suppose – but you can't go there! Quite apart from the appalling conditions it is the most dangerous spot in the whole of the Crimea. The Russians are constantly carrying out sorties, while their artillery is active for twenty-four hours of the day.'

'Then medical help will most certainly be needed there. Besides, that is most likely to be the place where Gwen will go and there are no doubt other wives up there too?'

'Probably,' Turnbull conceded, 'although I never saw any.'

'They would have been there, God bless 'em,' Nell declared, 'and they'll be glad of a little comfort, just as much as will their men. I'm here to see that they get it.'

In truth, Turnbull had visited the siege lines around Sebastopol only once, despite his ongoing and vivid reports to his London newspaper about conditions at 'the front'. Whenever he sallied forth from Balaklava in search of newsworthy material, it was usually either to the scene of a battle that had

174

already been fought, or to the headquarters of Lord Raglan, established at a safe distance from the regular exchange of artillery fire between besieged and besieger.

This did not prevent him from saying, 'Perhaps when you see something of the atrocious condition of the countryside, you will change your mind about moving away from Balaklava. In the meantime, I would be delighted should you decide to accept my offer of hospitality and move into the house I currently occupy. I took the house on the instruction of my employers who intend sending out photographers and more correspondents, but they will not be here in the foreseeable future. Until they arrive I am living alone in a house that is far too large for my needs.'

'It is a very kind offer, Mister Turnbull,' Nell said, 'but for the time being we will sleep on board with all my goods. During the day we will go ashore, see where our help is most needed and assess the general situation. I have no doubt we will meet up with you now and again.'

Disappointed, William Turnbull said, 'Well, if you change your mind – any of you – my offer remains open.'

This last remark was directed at Alice. When he left them, in order to make arrangements for himself and his belongings to be taken ashore, Nell said to her, 'Our Mister Turnbull fancies you almost as much as he fancies himself, Alice. Unless you want to encourage him I suggest you give him a wide berth. I've met too many Mister Turnbulls in my lifetime. They're trouble.' With a sudden grimace, she added, 'All the same, he might be useful to us. He knows Balaklava and the Crimea, we don't.'

'In the last letter I had from Gideon, he mentioned that the man he travelled here with was remaining behind to deal with the materials and equipment that should soon begin arriving for the railway,' Alice said thoughtfully. 'I think his name is Ranald . . . Ranald MacAllen. He probably knows the country between Balaklava and Sebastopol far better than Turnbull. He'll also know the route the railway will take. It might be useful for us to be close to the railway . . . for all sorts of reasons.'

'That makes a great deal of sense,' Nell agreed. 'You're more

175

than just a pretty face, girl. As soon as Turnbull's gone we'll all go ashore together. While I seek out one or two senior army officers for whom I have letters of recommendation, you see if you can find this friend of your young man. No doubt he can also tell you the best place to make enquiries about your husband, Gwen.'

The three women and Verity went ashore in one of the numerous small boats that dodged in and out among the many ships at anchor in the narrow seaway outside Balaklava. The Russian boatmen were making a great deal of money from the presence of their enemies, but Nell successfully negotiated a fare for the trip that was a mere fifth of that originally demanded. Despite this, she still complained that the Balaklava boatmen were 'a bunch of robbers'.

No sooner had the small party stepped ashore than an army major called out Nell's name and hurried across to embrace her warmly. Returning with him, Nell explained to Alice and Gwen that he had been one of her patients in Hong Kong, a few years before.

'I am proof – if any were needed – of the skills of Mother Harrup,' the major said, beaming at Nell. 'There are many others of the regiment who were in Hong Kong with me, who owe their lives to you. But what on earth are you doing here, in this God-forsaken hole?'

'Exactly the same as I was doing in Hong Kong,' Nell declared, 'but this time without the support of a husband.' Pointing to Alice, she added, 'This young lady has come to help me, while Gwen is here to rejoin her husband. She was on her way out here with him when their daughter contracted measles and they were put ashore at Malta. Alice is also hoping to meet her young man when he brings men out to build a railway for you.'

'There are two railway men here at this very moment,' the major said to Alice, 'the two men standing beside that large storehouse. I suppose one of them wouldn't be the young man you're looking for?'

'No,' Alice said, 'but I think the younger of the two is probably his friend Ranald MacAllen. Come along, Gwen, let's go and speak to him.'

V

Ranald MacAllen was talking to Robert Petrie but stopped in mid-sentence when he saw Alice approaching, evidently intent upon speaking to him.

'Excuse me, but are you Mister MacAllen . . . Ranald MacAllen?'

'Why, yes . . . I'm sorry, should I know you?'

Alice smiled at his confusion. 'No, but Gideon described you accurately in his letters, and when you were just pointed out to me as a railway builder, I knew it must be you. I am Alice Rowe.'

Ranald was startled. 'The girl from Cornwall? But . . . Gideon has returned to England expecting to find you there! How . . . what are you doing here in the Crimea?'

Alice looked momentarily unhappy. 'I was working for the Sisters of Mercy, in Plymouth, when they were asked to come and nurse at Scutari. They wanted me to come with them. I agreed because I thought Gideon was out here. It wasn't until I met a man named William Turnbull that I learned Gideon had gone back to England. In fact, I only missed seeing him by about an hour when his ship called at Scutari. Mister Turnbull told me Gideon would be coming back to the Crimea to build a railway, so when Nell Harrup asked me to come here with her I agreed, so I would be here when he arrived and not risk missing him again.'

'But what will you be doing here, in the Crimea? It's hardly the place for a young lady.' The question came from Robert Petrie.

Alice looked at him uncertainly and Ranald said hurriedly, 'This is Robert Petrie, Alice. He's one of the contractors responsible for building the railway here. In fact all three of us, Mister

177

Petrie, Gideon and myself, left London together. Unfortunately, Mister Petrie had a serious accident in France and had to stay in hospital for a while.'

'Yes, Gideon told me about it in one of his letters.' Sympathetically, Alice added, 'I hope you have fully recovered, sir.'

'I have, my dear – and I thought I remembered your face. I was with the British ambassador, in Paris, when we saw you and the other nurses off at the railway station. But I have seen none of Miss Nightingale's nurses in Balaklava – and who is this Nell Harrup?'

'She's a sort of nurse, although she prefers to call herself a "doctoress". She's earned a reputation all round the world for curing people – particularly soldiers – of all sorts of illnesses. She's skilled in dressing wounds, too. She's come here because she thinks she's needed. She also intends to open an eating-house to provide treats for the soldiers and their families. She asked me to come and help her.'

Looking mildly confused, Robert Petrie looked enquiringly at Gwen, who held Verity in her arms, and Alice said, 'I'm sorry, I should have introduced Gwen . . . and her daughter, Verity. Gwen's husband is in the Crimea with the army – the artillery. Gwen was travelling with him but Verity caught the measles and they were put ashore at Malta until she recovered. We travelled together from there to Scutari. They came here on the ship with Nell too.'

Nodding to Gwen, Robert Petrie returned his attention to Alice. Sounding dubious, he said, 'This project of Nell Harrup's sounds very ambitious. Where does she expect to obtain provisions in the Crimea for such a place?'

Gwen provided the answer. 'She's already got a shipload of stores and stuff for a restaurant and kitchen, out in the bay. She's being financed by a titled army officer she once cured of a serious fever. I believe more goods are on their way. She's hoping to buy anything else she needs here, providing she can get it at a reasonable price. She came ashore with us hoping to find a place where she can keep her stores until she finds somewhere to set herself up. At first it will probably be here, in

Balaklava, but her intention is to have an eating-house close to the siege lines around Sebastopol. That's where she says her doctoring skills will be most needed.'

Robert Petrie raised an eyebrow. 'I don't think anyone, military or civilian, would argue with that, but it will take a very special person to succeed in such a venture. I would like to meet this lady. We might possibly come to an arrangement that could work out to our mutual advantage.'

Ranald looked at the contractor quizzically and Petrie explained, 'The warehouse we have taken over will be empty until our equipment starts to arrive – which won't be for at least another month. If this lady is genuine she could make use of it until then. Do you think you could arrange for me to meet Nell Harrup, Alice?'

A few minutes later Alice and Gwen, with Verity, were hurrying off to find Nell and tell her the exciting news.

They found her on the dockside, angrily puffing at a cheroot. Before they had an opportunity to tell her their news, she said, 'That damned fool of a harbourmaster says there's no chance of bringing the *Princess* inside the harbour for at least another week – and then he won't guarantee that we'll be able to berth alongside. The ship's master won't keep the *Princess* idle for all that time. He's likely to take the cargo back to Scutari and offload it there. If he does, it'll be stolen long before we can find another ship to bring it back here again – but I'll show him! I'll hire every boat in Balaklava to bring my goods ashore. It'll clog the harbour and throw his system into chaos.'

'Before you do that, go and speak to Mister Petrie.' Alice spoke placatingly. Telling her what the railway contractor had said, she added, 'Gideon told me he's a very good man – and, because of the importance of the railway he's going to build, he has a lot of influence with the authorities here.'

Nell was not entirely convinced, but she went with the others to meet Petrie and it was immediately apparent to Alice that she and the contractor were two of a kind. Both were used to getting things done, and neither would accept failure if there was something they really wanted.

179

After a lengthy discussion, Petrie agreed he would take Nell along the proposed railway route, in order for her to choose a suitable site for her eating-house. In return, her skills as a 'doctoress' would be at the disposal of the navvies. They would be given precedence when she began catering for the needs of those engaged in the war.

Once this had been agreed, Petrie was able to provide Nell with the answer to her current problems. He possessed a document signed by the British war minister, giving his requirements priority over all but on-the-spot battle orders.

Invoking this document, Petrie would ensure that the *Princess* was brought alongside almost immediately in order that unloading might commence.

For the moment, Gwen was less fortunate. There was no one in Balaklava who could tell her where she might find her husband. There were a couple of artillery posts just beyond the Balaklava pass, but she was told it was doubtful whether he would be there. He was most likely with the main body of the artillery, involved in the siege of Sebastopol and in daily action against the Russian troops defending the Crimean sea port.

But Petrie was able to help here, too. He had spent time at Lord Raglan's headquarters, discussing how best the railway could serve the British army, particularly the artillery, who were having considerable problems transporting ammunition to the siege lines for their guns. As a result he had come to know a number of artillery officers. He promised to write a note to one of the most senior, enquiring after Gwen's husband.

Robert Petrie and Ranald were also able to offer the three newly arrived women accommodation in the house that had been taken by Gideon and Ranald when they first arrived in Balaklava.

As a result, that night Alice slept in the bed that had been occupied by Gideon during his stay in Balaklava. A few of his belongings were still in the room and they gave her a strange and intimate sensation. It seemed to her she could almost feel his presence close by.

Before going to bed, she wrote him a letter, telling him of the change in her circumstances. Ranald had promised her it would go out with the mail he and Petrie were sending the next morning, although it was by no means certain it would reach Gideon before he left England with his navvies.

VI

The *Princess* was brought alongside a harbour quay at first light the next morning and, early though it was, Nell was there to ensure that unloading was begun immediately by a team of Turkish labourers.

The men were supervised by 'Ali', also a Turk, brought to Balaklava from Scutari by Nell. The doctoress had a happy knack of attracting men to her who had the qualities she demanded and Ali in particular was to prove indispensable.

Ali was not his real name and when Nell first employed him he objected strongly to her use of the simple cognomen. However, she told him she found his given name totally unpronounceable, adding that she had no intention of learning a whole new language merely in order to be able to pronounce the name of an employee, so 'Ali' he became.

The Turk possessed many qualities, not least of which was an uncanny ability to detect a thief, no matter how clever he tried to be.

Alice witnessed this skill later that morning when she made her way to the dockside. Ali had just caught a young labourer who had managed to wrap an incredible length of light rope about his body beneath his loose caftan. He was now cuffing the unfortunate thief about the head as he drove him from the quayside.

When she commented to Nell upon Ali's success, the healer merely shrugged. 'You know what they say, Alice, you set a thief to catch a thief – especially if he's better at it than they are.'

'Ali is a thief?' Alice was aghast. 'Aren't you afraid he'll steal from you?'

'He might,' Nell agreed nonchalantly, 'but he'll earn his money by stopping everyone else doing it. Besides, he won't take enough to be noticed. I've told him that if I so much as catch him taking a bite from an apple I'll send him back to Scutari. That's enough to put the fear of God into him.'

'Why, what's he done there?' Alice asked.

'I don't know,' Nell admitted. 'But the day after he started working for me he came to me in a great state asking to be allowed to sleep on board the *Princess*. He never showed his face above deck again until we left harbour.' Nell grinned broadly at Alice's expression of concern. 'I don't think we need worry too much about whatever it was he did, Alice. It probably had something to do with his partiality to alcohol and other men's wives. Neither trait is likely to endear him to his own people. He's a rogue – but a likeable one. Now, have you any idea where Gwen is at the moment? Robert Petrie gave me a message for her.'

'She washed some of Verity's clothes last night and left them to dry around the fire. When I left her she was trying to borrow an iron to press them.'

'She needn't bother. They'll only get creased again when they're packed. It seems Lord Raglan has decided he now wants the railway line taken up to each of the artillery positions along the front line to guarantee them a continuous supply of shells. Petrie and MacAllen are going to Raglan's headquarters this afternoon, then on to the artillery posts tomorrow to see what will be needed. They'll have an escort and Petrie says he'll arrange for an extra horse and take Gwen and Verity with him. Her husband is up there somewhere and they're bound to find him.'

'Gwen is going to be absolutely thrilled,' said a delighted Alice. 'Do you need me for anything here, Nell, or can I run back to the house and tell her the news?'

'There's nothing you can do here. Go and tell her – then help her pack and get Verity ready. She'll be too excited to do anything properly. I don't know why we let our men get us into such a state, they're none of 'em worth it in the long run.' Still grumbling good-naturedly, Nell added, 'We'll no doubt have you running around in the same state when your man arrives in a few weeks' time.'

She broke off to utter dire threats to two Turkish labourers who were not handling a crate containing crockery with sufficient care.

Hurrying to the house where she had left Gwen, Alice thought of what Nell had said about Gideon arriving in the next few weeks. It gave her a nervous thrill of anticipation. She and Gideon had spent such little time together. She wondered whether he might have had second thoughts about her, especially as he was now far more important than he had been when they first met.

Gwen was just as excited as Alice had predicted she would be. Unable to decide what she and Verity should wear for the ride to the army headquarters and what they would save for the following day, she packed and repacked their clothes as she repeated to Verity that she would 'be seeing Daddy tomorrow'.

Alice cooked a midday meal aware that Gwen was likely to eat very little, but she included enough for Nell. As she had anticipated, the cockney doctoress came back to the house to bid farewell to Gwen and Verity. She brought a bag of provisions for Gwen to take with her, selected with all the experience of her many years as an army wife.

Nell returned to the dockside immediately after the meal, promising to visit Gwen and Verity in the siege lines in the very near future. She passed Robert Petrie and Ranald MacAllen in the street outside the house and a few minutes later Gwen was mounting the spare horse they had brought with them. Verity would ride with a trooper of the escort who would take her with him on his horse.

Knowing how much Gwen had pined for her husband, Alice was happy that she would soon be reunited with him, but she would miss her – and the small girl too. Verity was a warm and loving child and Alice had to try hard not to shed a tear as she kissed her goodbye and passed her up to the trooper.

'You take very good care of her,' she said. 'She's a very special little girl.'

'Of course she is, miss,' replied the cavalryman, settling Verity on a folded blanket on the saddle in front of him, 'and you don't need to worry yourself about her. I have one just like her at home and can't wait for the day when I'm back with her and her mother.'

Alice waved until the small mounted party was out of sight before picking her way to the harbour along the muddy roadway. She was startled when a voice called her name, and turned to see William Turnbull hurrying to catch up with her. She did not slacken her pace and had reached the road that led to the harbour by the time he caught up.

Breathing heavily from his exertions, Turnbull wheezed, 'You set quite a pace, Alice. Tell me, was that Gwen and her little girl I just saw riding off with Petrie, MacAllen and an escort?'

'That's right. She's going up to the front to be reunited with her husband.'

'She'll be a very happy lady – but how about you? You're going to miss her.'

'Nell will find plenty for me to do,' Alice said shortly.

'Well don't let her work you too hard. You deserve some time off after all the work you did at Scutari. We'll have to get together, Alice. In spite of the war and all that's going on here, there are some quite pleasant places not very far away. I'll hire a couple of horses and show you.'

When Alice made no reply, Turnbull persisted, 'We still have to take a photograph of you for my paper and I could write an article about your being the first British nurse to actually come to the Crimea to tend our troops. Your fame would be such that you'd be fêted everywhere you went when you returned to England.'

'I'm quite happy with what I'm doing at the moment, Mister Turnbull. I have no wish to be famous – what's more, I'm not even a nurse.'

184

'That makes your achievements all the more remarkable. Think about it, Alice. You could return home and earn a very comfortable living travelling round the country talking of your experiences here.' When Alice said nothing, the reporter went on, 'I could do that for you, Alice – and I would be much happier if you were to call me William, and not Mister Turnbull . . . but here comes Mrs Harrup. I will leave you, but I will speak to you again soon. Goodbye, my dear.'

With a waved acknowledgement to Nell Harrup, the correspondent turned and hurried away and Alice smiled. She did not like William Turnbull and never felt comfortable in his presence, but she knew she was quite safe as long as Nell was around. For some reason the correspondent feared her far more than he did any of the senior army officers.

14

I

So many applications were received from navvies eager to work on the Crimea railway, Gideon could have engaged a work force ten times the size of the one needed. Recruiting was completed in just two days, instead of the week that had been set aside for the task.

As a result it was decided to bring forward the sailing date for the railway builders. On the day the ship carrying Alice's letter from Balaklava docked in Falmouth, a clipper, the *Argo*, set sail from Liverpool with Gideon, an advance party of navvies, and a large quantity of tools and equipment.

The departure came as a relief to many of the city's publicans. Among those taking passage on the *Argo* was a full gang of Welsh navvies. Most were ex-coalminers and they were led by their own ganger, a big, black-bearded man named Bryn Roughley.

They had been waiting in Liverpool for a week, becoming increasingly bored and aggressive. The night before the *Argo* sailed the whole gang went out on a randy. The ensuing brawls involving navvies, dockers and police would be talked about for many months.

Despite their drunken carousing, every one of the Welsh navvies was on board the ship when it set sail on its long journey bound for the Crimea, although the faces of many bore the scars of battle.

Gideon's own gang were on the ship too. He and Sailor stood on the deck as a steam-tug eased the ship from the dockside. Listening to the taunts and insults hurled at the dockworkers by the Welsh navvies, who were convinced they had proved their superiority the previous evening, Sailor commented, 'It's going to be a long voyage, Gideon. That lot are trouble.'

'I've never yet come across a worthwhile gang that wasn't trouble in one way or another, Sailor. Once we're in the Crimea they'll get the job done well enough.'

'I don't doubt it,' Sailor agreed. 'But there's a whole lot of water between Liverpool and the Crimea. I hope most of it's rough enough to give them something more than trouble to think about.'

The weather on the ten-day voyage between Liverpool and Gibraltar might have been arranged to meet Sailor's wishes. It was a mixture of rough and smooth sailing. Most days were rough enough to make the men want to spend more time lying on their bunks than prowling round the ship becoming bored. Such periods were interspersed with relatively calm days which gave the navvies just enough time to recover from their seasickness before the weather changed once again.

On the tenth day the weather relented, enabling the navvies to come on deck and be suitably impressed by the towering Rock of Gibraltar as the *Argo* sailed into the harbour of the tiny British colony whose fortifications guarded the gateway to the Mediterranean. Gideon watched with mixed feelings as they trooped ashore, eager to sample all that the colony had to offer them.

Sailor too had misgivings. 'We could have done with the weather being a lot rougher these last couple of days, Gideon. The men have too much energy left.'

Gideon smiled at these words from a man who only six months before would have led his fellow navvies ashore and downed his first tankard of ale before the last of them had stepped from the gangway.

'The responsibility of being a ganger has tamed you, Sailor. I'd never have believed it could happen.'

Looking only slightly abashed, Sailor replied, 'Seeing what *you've* achieved is what's done it, Gideon. I remember when you first became a navvy. I knew then you were a lot brighter than most, but I could never have imagined just how far you would go in only a few years. I can't see myself ever becoming a contractor's right-hand man, but standing in for you has made me realise I *could* become a ganger.'

'Good!' Gideon said sincerely. 'When we return from the Crimea I should be able to sub-contract, at least – and you'll be my first choice as ganger.'

'I'll remind you of that when the time comes,' Sailor said, 'but before we get all carried away with the future we'd better make certain of the present. That means keeping this lot in order. Are you going ashore?'

Gideon nodded. 'It might be an idea if we go together. I don't anticipate too much trouble with our own men, but I am concerned about the Welshmen. They're even wilder than our own.'

Gibraltar was a very busy port of call for ships of all nations. Taverns catering for the whims of the sailors were very much in evidence along the main street of the tiny British outpost, with signboards advertising the delights of each, written in many languages, displayed outside their open doors. Standing beside the doorways, men and women called, cajoled and occasionally bullied men from the ships into entering their particular establishment to sample what was on offer.

One of the largest of the taverns was particularly noisy. It was here, later in the evening, that Gideon and Sailor found the Welsh navvies.

There was a show, of sorts, taking place on a flimsy wooden

stage at the far end of the tavern. Here a once-young Spanish woman held castanets at arm's length above her head in an attempt to perform a gypsy dance to the music of a trio comprising drummer, trumpeter and guitarist.

Of the three instruments, the trumpet was by far the loudest, but acute nervousness was causing the player to produce notes that would have been unfamiliar to the music's composer.

The source of the discomfiture of both trumpeter and dancer was the Welsh ganger, Bryn Roughley, who was taking an uninvited part in the performance on stage. With hands held to the side of his head, forefingers extended in a manner intended to represent horns, he was pretending to be a bull, charging the gypsy dancer and forcing her to keep sidestepping in order to avoid being knocked from the stage.

Eventually, as Gideon and Sailor watched, the dancer, exasperated with the Welsh 'bull', spat out a string of invective at him. Then, lifting her frothy, layered skirt clear of the dusty boards, she gave a contemptuous flick of her head and stalked from the stage. As she disappeared behind a long and grubby curtain that hid the performers' tiny changing rooms, the strident notes of the trumpet faltered, then ceased altogether.

At this, such a howl rose from the rowdy navvies that the nervous trumpeter decided to follow the dancer's example and make his exit. Scrambling on to the stage he followed the route taken by the dancer – but as he did so the trumpet was seized from his hand by one of the navvies. For a moment it seemed he might contest the ownership of his brass trumpet, but sanity prevailed. He fled across the stage, closely followed by his fellow musicians, only the guitarist retaining possession of his instrument.

There was a half-hearted move to follow the departing trio, but then the Welshman with the trumpet put it to his lips and produced a sound which was as loud as it was unmusical. At the same time, another of the navvies picked up the drumsticks from the floor and began attacking the drum with them.

It was the signal for most of the other navvies to take to the boards. Climbing on to the stage, they each performed a highly

individualistic version of the dance performed under such difficult circumstances by the gypsy dancer.

The stage had not been constructed to withstand the stamping
feet of so many muscular navvies. It collapsed, the sound of
splintering wood adding to the noise of trumpet, drums and
shouting men.

'We'd better get this lot out of here,' Gideon said. 'There's
going to be trouble.'

'Too late,' said Sailor. 'I think trouble has arrived.'

The two men were standing in the doorway and Sailor had
heard the shout go up for help and the sound of running feet
in the street outside. A number of uniformed policemen were
pelting along the street towards the noisy tavern and Gideon
and Sailor stepped aside as the first of them arrived. For just a
moment one of the policemen paused to look at them and
Gideon thought they were about to be taken into custody.

A great shout from inside was accompanied by the sound of
breaking glass. Glancing into the tavern, Gideon saw the
navvies helping themselves from the bar, while frightened
bartenders cowered beneath the long wooden counter. Then he
and Sailor were pushed aside as more policemen reached the
scene. The new arrivals paused just inside the tavern and
waited until they felt they were in sufficient numbers to restore
order.

What followed was a pitched battle. The policemen used
their staves freely in a vain attempt to subdue the Welsh
navvies, who, in a fighting defence, made use of anything that
came to hand, their weaponry including bottles, chairs, and
table legs.

Gideon went inside in a bid to stop the fighting, but coming
under attack from a policeman he retired to the doorway and
looked on helplessly as the pitched battle continued to rage.

More and more policemen appeared on the scene, but the
Welshmen only fought the harder. Then a senior officer arrived
and Gideon prevailed upon him to call off his men. He had been
reluctant to comply at first but Gideon managed to convince
him that he had the authority of the British government behind

him. The senior officer finally agreed, but told Gideon the chief constable would want to talk to him before the ship left Gibraltar.

Not until the policemen withdrew, with a mixture of relief and reluctance, was order restored once more. The Welshmen felt they had secured a victory and were cheering their success when Gideon called for their attention.

When he had it, he said, 'Right, you've had your fun, now you can all return to the *Argo* and we'll talk about this in the morning.'

'What is there to talk about?' Bryn Roughley demanded. 'We've had a bit of a fight, but now the police have withdrawn. We've come ashore to enjoy ourselves, and that's just what we're going to do.'

'The police have withdrawn because I asked their senior officer to call them off. If I hadn't they'd have called out the army and you'd all have ended up in cells, appeared in court in the morning, and been sent to prison. I've no doubt you all thoroughly deserve it – but my job is to get you to the Crimea and build a railway. So go back to the ship. I'll see you in the morning. If you don't, I'll have to go outside and tell the police chief there's nothing more I can do. He'll send for the army and, sooner or later, you'll all be arrested and I'll need to send for replacements for you. It's up to you. As I said, you've had your fun. Now it's time to go back on board the *Argo*, while I try to sort things out here.'

Some of the men seemed disinclined to heed Gideon's warning, but after apparently weighing up the situation Bryn Roughley said, 'All right, Mister Davey, we'll do as you suggest – this time – but you're employing us to work for you, not to tell us how to spend our time when we're not working.'

'All I'm interested in is getting you to the Crimea where the work is,' Gideon retorted. 'Once you've done the job you're being paid for you can do what the hell you like. But a great many lives – British soldiers' lives – depend on you getting out there and building a railway for them. My job is to see you do it. If you can't then I'll get men out here who can. Now, get your men back to the ship before the police chief changes his mind and arrests you all anyway.'

II

The morning after the fracas in the Gibraltar tavern, Gideon was summoned to the colony's police headquarters. Here he spent an hour talking to the chief constable who, fortunately, was impressed by the sound of the work on which the navvies were to be employed in the Crimea. Nevertheless, he was determined not to be completely overawed by the seriousness of the task to which Gideon was committed.

'You have convinced me that your work in the Crimea is of vital importance, Mister Davey. Had you not done so, and had the national interest not been involved, I would have demanded that you hand over every one of the men involved in the disgraceful affair of last evening and personally ensured that they languished in a Gibraltar gaol for a very long time.'

'I am aware of that, sir, and were I in your place I would feel very much the same,' Gideon said apologetically, 'but these men are irreplaceable. By their efforts they will save the lives of hundreds – if not thousands – of British soldiers. I regret the damage to the tavern, but if the owner will give me an estimate of the cost incurred, I will write a banker's order to reimburse him.'

'I have his estimate here,' the chief constable said, pushing a piece of paper across the desk that separated the two men. 'It is, of course, grossly exaggerated, but if you write a banker's order for half the amount, it should close the matter.' Giving Gideon a look that contained a hint of sympathy, he continued, 'I suggest you warn your navvies against behaving in a similar manner when they reach Malta, Mister Davey. The military are policing the island at the moment and they are far less understanding than we are here in Gibraltar.'

'I will bear that in mind,' Gideon replied. Aware that the

policeman could have made things a lot more difficult for him, he added, 'And I would like to thank you for your co-operation and understanding.'

Back on board the *Argo*, Gideon called all the Welsh navvies together on the upper deck. They were in high spirits, even though the bruised and battered faces were evidence that their battle with the Gibraltarian police had not been entirely one-sided.

Gideon needed to call for silence a number of times before the navvies heeded him. When he had their attention, he said, 'I have just returned from a meeting with the chief constable of Gibraltar. Needless to say, I didn't enjoy it.'

There were a number of comments in response to this and one navvy called out, 'Then mind your own business and we'll sort out our own randies. We don't need to be wet-nursed by anyone.'

There was general approval for this comment and Gideon retorted, 'If I hadn't been there to "wet-nurse" you there'd have been no Welsh navvies on board the *Argo* when she sails from Gibraltar today. Those of you who weren't arrested last night would have been taken off this morning. The chief constable told me the ship wouldn't have been allowed to leave harbour until you'd been handed over.'

The muttered dissension was only a little more muted and Gideon said, 'Fortunately I was able to convince him of the importance of what we have to do in the Crimea. After I paid for the damage that was caused to the tavern last night, he agreed not to pursue the matter.'

His words brought a chorus of derisive cheers from the Welshmen. When it had died down to no more than an amused murmur, Gideon said, 'I wrote a banker's order for seventy-five pounds, half of what the tavern owner claimed . . .'

Once again Gideon was shouted down with howls of derision and disbelief, and he waited patiently until he could be heard again.

'I don't intend that the company should pay for the results of your randy. There are fifty of you, so you'll all be docked one pound ten shillings from your first month's pay when we reach the Crimea.'

This time the howls were of fury and one man shouted, 'That's not fair. I didn't even go ashore last night!'

'Then I suggest you persuade the others to have a whip-round to make it up to you,' Gideon retorted. 'That's all I have to say for now. The ship will be sailing in an hour's time and the next port of call is Malta. I've been told the army carry out the policing there, so I suggest you stay out of trouble, unless you want to end up with a flogging.'

Leaving the men angrily discussing what he had just said to them, Gideon headed for a companionway that led to the cabins on a lower deck. At the top of the steps, Bryn Roughley stepped into his path.

'I think you and I should have a talk, Mister Davey.' Roughley glowered at Gideon. 'The men aren't happy with the way you handled things on shore, and neither am I. We're contracted to Mister Petrie, not to you. You may be his agent, but I'm still their ganger. You had no right to dock their pay, no right at all – and certainly not without discussing it with me.'

'And when do you suggest I should have discussed it with you?' Gideon retorted. 'Would you have liked me to have sent for you this morning when I was apologising to the chief constable for the behaviour of you and your men? Or perhaps you think I should have climbed on to the stage to speak to you last night, when you thought you were a bull?'

'We were out on a randy. You know how it is. At least, you did until you caught Petrie's eye.'

'I know exactly how it is, but now I've "caught Petrie's eye", as you put it, I've been given a job to do. It's probably the most important that any of us have ever had to tackle and it won't get done if I lose fifty men whenever the *Argo* puts into port.'

'It wouldn't have come to that . . .'

Roughley spoke scornfully, but Gideon cut him short. 'It damn near did! If I hadn't convinced him that I had letters from some very important people to back up what I told him, the chief constable would have stopped the *Argo* from sailing until you and your men had been taken off the ship and handed over to him.'

Bryn Roughley was not convinced and he said so. 'You behaved high-handedly, Mister Davey. You need my men on your side, and you won't get the best out of us by docking our pay before we've even lifted a shovel.'

'This morning I had to pay out seventy-five pounds of Mister Petrie's money for you and your gang. He's paying you a better wage than you've ever earned before. I see no reason why he should have to cover the cost of your randy too. And you and your men had better remember what I've said about your behaviour in Malta. The army would just as soon flog a navvy as a soldier.'

III

Between Gibraltar and Malta the smouldering resentment harboured by the Welsh navvies against Gideon twice erupted in fighting between Bryn Roughley's navvies and Gideon's own men.

On the second occasion, although Roughley had been instrumental in stirring up the trouble, he had not taken part in the ensuing fight which ended when Sailor stepped in and felled two of the Welsh ringleaders. It was then that Roughley appeared on the scene and ordered his men back to their quarters.

The navvies reluctantly did as they were told, but as they went off nursing their grievances and bruises Roughley rounded upon Sailor. 'There was no need for you to become involved. You should have let the men sort things out for themselves.'

'Your men are spoiling for a fight,' Sailor said, 'and have been since we left Gibraltar. The two I downed were egging the others on. If I hadn't stepped in they might well have caused a riot on board. Had that happened, Gideon would have been left with no alternative but to have your men taken into custody and sent back to England when we reached Malta.'

'He wouldn't have done that. He needs us in the Crimea,' Roughley said smugly.

'He needs a work force he can rely on to do the job they're being paid for, not waste time fighting each other.'

'Well, you've done nothing to end the bad feeling between them. You've knocked down two of the gang and they'll want to get their own back, unless . . .' Roughley paused and looked at Sailor speculatively.

'Unless what?' Sailor prompted.

'Perhaps we can each put up a champion in a prize-fight and have a purse of, say, seventy-five pounds? That would repay the money taken from the men by Davey, and satisfy the pride of my gang.'

'You sound very certain your man would win,' Sailor said.

'Oh, it would be a fair fight – but we have better men,' Roughley said arrogantly.

'I admire your confidence, but even if I agreed I doubt whether the captain would allow us to stage a prize-fight on his ship.'

'I wasn't thinking of having it on board. It could be arranged when we reach Malta.'

'Gideon has already told you that the army's policing Malta. They'll arrest anyone they find even *watching* a prize-fight. Your men might not be such a loss to the Crimea, but if my men were arrested too it would be a disaster.'

Ignoring the insult, Roughley said, 'There must be somewhere on the island to stage a prize-fight. The army can't get everywhere.'

An idea was forming in Sailor's mind as the Welshman spoke. Suddenly thoughtful, he said, 'There *is* a place . . . In my Navy days I once took a boat up Sliema Creek. It was pretty quiet there . . . but if I *did* stick my neck out and agree to this idea of yours I'd need to be certain it would clear the air between our gangs once and for all. I don't think we could do that by choosing just anyone from each of our gangs.'

'What are you suggesting?' Roughley asked suspiciously.

'I'm suggesting the fight is between you and me. It would carry far more weight with the men, especially if we let it be known it's to clear the air between the gangs once and for all, regardless of who wins or loses.'

The suggestion took Roughley by surprise and he looked at Sailor speculatively. The acting ganger was a big man, but the Welshman had fought and beaten big men before. They were also much of an age.

'What about the purse?' he asked.

'We'll have no problem putting up money,' Sailor replied. 'But we'd want to see the colour of your money too, no matter how confident you are of winning. The most important thing is to make it clear that there is to be no more trouble between our gangs after the fight. We can only do that if you and I set an example to them.'

Roughley thought about what Sailor had said, then nodded his head. 'All right, I agree . . . there's my hand on it.'

The two men shook hands, and despite a lack of warmth in the handshake each knew the other would keep his word. Win or lose, the open animosity between the two gangs would come to an end.

News of the planned prize-fight caused a great deal of excitement among the navvies, especially in the gangs of the two men involved. They began to talk to each other – albeit only to lay bets on their respective gangers.

Gideon noticed the apparent lessening of the animosity between the two groups of men and commented upon it to Sailor. Tongue in cheek, Sailor replied that he had always believed it would happen once the men accepted they would all be working together in the Crimea.

Six days after setting sail from Gibraltar, the *Argo* sailed into the harbour at Valetta.

Once inside the towering walls, which were topped with centuries-old houses and fortifications and cut off the wind that was the sole means of power for the large sailing vessel, the *Argo* was becalmed until a small, noisy tug took her in tow and guided her to a suitable anchorage.

The ship's captain announced that the passengers could go ashore as they wished, but they would need to make their own arrangements to make the short crossing. This posed no problem. Numerous local boats were gathering in rapidly increasing numbers round the latest arrival in the harbour, boatmen underbidding each other as they clamoured for business.

Gideon spotted Sailor among those waiting at the top of the ship's gangway to be one of the first ashore.

'I'm surprised to see you in such a hurry to leave the *Argo*, Sailor. I was going to invite you to come ashore and have a couple of drinks with me later.'

'You know me, Gideon, a much travelled man with friends in every port. I spent time here during my Navy days. I'm going ashore to meet up with someone – perhaps we can have that drink together later?'

There was something in Sailor's manner that Gideon found vaguely disturbing, but watching the ex-Navy man heading for the shore in a Maltese boat he tried to convince himself he was imagining things. Sailor had led a full life before becoming a navvy. He was probably going ashore to meet up with a girl he had once known here.

Only minutes after Sailor had departed, Bryn Roughley came up on deck with members of his gang, all bent on going ashore. Before the Welsh ganger left the ship Gideon called him to one side and gave him a stern warning about the conduct of him and his men while they were in Malta.

'You don't need to worry about us, Mister Davey,' said Roughley cheerfully. 'We'll be so well behaved no one will even know we're here.'

Such politeness worried Gideon more than Roughley's previous belligerence, but there was little he could do except watch the Welsh ganger and his men being rowed ashore and hope they would not get into any more trouble.

IV

In Valetta, Sailor quickly moved out of sight of the *Argo*, leaving one of the navvies to direct the remainder of the gang and the Welshmen to a side street where he would be waiting. It was another twenty minutes before both gangs were gathered together and Sailor started to lead the way towards a less populated part of the island.

Things had changed since Sailor had last visited Malta. He was somewhat disturbed to find there was now a Royal Navy stores vessel anchored in Sliema Creek. There were a number of new houses in the area, too, and the occupants looked out of their windows apprehensively as more than a hundred navvies passed by. But there were open spaces here as well and soon a spot was found, overlooked by only a couple of Maltese-occupied houses, which both combatants agreed would be suitable for their prize-fight.

As the contestants stripped to the waist, the navvies formed a large square within which the fight would take place. Each man had two 'seconds' armed with towels, water and brandy, to bathe battered faces and revive the fighters when necessary. There was an air of great excitement among the spectators, not least because a considerable amount of money had been wagered on the outcome of the contest.

The rules, such as they were, were simple. It was to be a 'no-holds-barred' contest. Neither man was expected to deliberately set out to cause his opponent permanent injury, although, if it occurred, it was an accepted hazard. Fists were the main weapons in the contest, but feet, head and wrestling holds were not barred. A round would last for as long as both men remained on their feet. When one was knocked down and unable to regain his feet immediately, a one-minute rest period would be called. If the fallen man was unable to return to the fray after this time, the other man would be declared the winner. The fight would continue until this point was reached.

Sailor expected it to be a lengthy contest. Indeed, he planned that it would be. Roughley was known to a number of navvies outside his own gang. By talking to them, Sailor had learned that the Welshman had been a miner for many years before taking work on the railways. As a result, his lungs gave him trouble when subjected to periods of sustained exertion. His problems had eased when he became a ganger because, unlike Sailor who worked alongside his men, Roughley preferred a purely supervisory role.

For this reason, Sailor knew Roughley would try to bring the contest to a swift conclusion. An experienced fighter, Roughley was known to use his heavy boots to hasten the end of a fight. But Sailor had also noticed that the Welshman tended to favour his right leg and occasionally moved with a noticeable limp. If need be, Sailor would use this knowledge to his advantage.

The fight began quietly enough with both men circling each other for almost a full minute, occasionally jabbing out a fist, seemingly more to keep the other man at bay rather than to inflict injury upon him. Then, when the watching navvies were beginning to call for more action, Roughley leaped forward and, feinting with his arm, aimed a savage kick at Sailor's knee.

Fortunately, Sailor had been expecting just such a move and the Welshman received a painful fist in his left eye.

The fight became more tense now, with each man manoeuvring to gain an advantage over the other. For more than four minutes they jabbed and ducked and exchanged an occasional flurry of punches. Once too, when they closed on each other, Roughley succeeded in scraping Sailor's shin with the heel of one of his heavy boots. The sudden pain brought swift retaliation. Landing a number of heavy punches, Sailor knocked his opponent to the ground.

Roughley was not seriously hurt, but sensibly decided to remain on the ground long enough for a rest to be called.

'That shook him up,' said the navvy wiping Sailor's face with the towel. 'You've got the beating of him, Sailor.'

'Don't underestimate him,' Sailor said in return. 'His boot on my shin probably hurt me more than my punches hurt him. Bryn Roughley is going to be a tough man to beat.'

The next twenty minutes proved the accuracy of Sailor's assessment of Roughley's capabilities. The Welshman continued to fight with ferocity and determination – but as the bout progressed Sailor sensed a growing desperation in the other man's attacks. He believed Roughley was becoming tired.

But Sailor too was slowing and when the other man came forward, swinging wildly, he failed to avoid a low punch that caught him in the groin.

Gasping painfully, Sailor dropped to one knee. It was the opportunity for which Roughley had been waiting. He kicked out at his opponent, and although Sailor threw his head back the boot struck his eyebrow and laid it open, blood gushing down and blinding him in the left eye.

Roughley aimed another kick intended to end the contest, but with his one good eye Sailor saw it coming and grabbed the foot with both hands before it struck. This was the leg he had noticed the Welsh ganger nursing on occasions and he twisted it cruelly.

Roughley cried out as a fierce pain shot through his knee joint. It was his turn to drop to the ground, his face contorted with agony. Nevertheless, he rose to his feet surprisingly quickly, aware that he needed to end the fight now if he was to stand any chance of winning – but the opportunity had gone.

Sailor too realised that he must end the fight quickly. Cuffing blood from his blinded eye, he flung a barrage of heavy punches at Roughley, moving round him as he did so. The Welsh ganger's injured knee prevented him from turning quickly enough to ward off Sailor's sustained attack and he took severe punishment before he was struck by two mighty blows, the first beneath his ear and the second flush on the point of the jaw.

Roughley dropped to the ground and a break was called. His two seconds bent over him where he lay, pouring water on his face, slapping his cheeks and calling for their man to rouse himself and continue the fight.

Their efforts and the urgent shouts of the navvies of his watching gang were all in vain. By the time the minute was up, Roughley was only just beginning to make low, moaning sounds. He was in no condition to continue. The fight was over. Sailor had won. His gang were jubilant, making so much noise that the alarmed cries of some of the men on the edge of the crowd of spectators went unheeded until it was too late.

The warning was of a large body of soldiers, running towards them. Holding a towel to his badly cut eyebrow, Sailor saw them coming and shouted, 'Scatter! Everyone run for it. Get back to the ship – and Roughley's gangers . . . take him with you.'

The navvies began running in all directions and the soldiers faltered uncertainly for a few minutes. Then, at an order from their officer, they too broke formation and began individual pursuits.

Sailor dropped down to the edge of the creek and, ducking out of sight, began running, a number of his navvies following. His last glimpse of Bryn Roughley showed the Welshman being dragged between two of his men, still half dazed and hopping painfully in an effort to maintain his balance while many of the Welsh navvies ran to his aid, determined to ward off the soldiers.

V

Called to the *Argo*'s sickbay by a surgeon who had been employed to travel to the Crimea with the navvies, Gideon found Sailor having his split eyebrow stitched together.

'What's been going on?' he demanded. 'How did you get that?'

Reluctantly, Sailor admitted, 'I got it in a prize-fight. I was fighting Bryn Roughley.'

'You were fighting . . . ! What the hell did you think you were doing, Sailor? I've had more than enough trouble with Roughley and his gang on this trip without you following his example. Where's Roughley now?'

Wincing as another stitch was put in the eyebrow, Sailor

replied, 'I don't know, Gideon. I . . . I think he's probably been arrested. He and many of his gang. The army put in an appearance just after the fight had ended.'

Gideon groaned. 'That's all I need! Just what did you think you were doing . . . you of all people? I thought you were the one man I could rely on.'

'The fight was intended to clear the air between our two gangs. Roughley and me shook hands on it. Whatever the result, we agreed our differences would be over . . .' Sailor told Gideon of the conversation between Roughley and himself.

'Did you really think a fight would settle all the differences between our gangs, or was it just an excuse for Roughley and his men to stir up more trouble for me?'

'It was my idea, not his. There'd have been no more trouble between us, I'm sure of that. Bryn Roughley gave me his word.'

'And you believed him? You didn't think things would be even worse once you'd beaten him – as I don't doubt for a moment that you did?'

'If I hadn't believed it I wouldn't have fought him,' Sailor said. 'He meant it, Gideon. I think he was as keen as we were to get things sorted out. His gang is one of the best . . . but you don't need me to tell you that. You knew of their reputation when you took 'em on.'

'Have any of his gang returned to the ship?' Gideon demanded.

'I don't know,' Sailor replied unhappily. 'When I came on board I came straight here.'

'Well, you're sorted out now,' commented the surgeon, standing back and looking critically at his handiwork. 'But you'll always carry a scar to remind you of Malta.'

'In that case you can go to the Welshmen's quarters and find out how many men we have left and whether anyone saw exactly what happened to Roughley and the rest of his gang.'

Gideon spoke angrily. This was exactly the sort of trouble he had hoped to avoid. The chief constable at Gibraltar had been both understanding and co-operative. He doubted whether the commanding officer of the Maltese garrison's police section would prove as helpful.

Gideon's forebodings were well founded. Major Courtney Archer was provost marshal for the islands of Malta. Although the post on the staff of the garrison's commander brought him prestige as chief of the army's police, it was one he did not enjoy – and it carried with it very little hope of advancement.

An ambitious man, Archer had married the younger daughter of a general with a view to furthering his career. It had not helped. He had needed to prevail upon members of his own family to lend him the money with which to purchase promotion to his present rank. Now, commander of officers whose regiments had been eager to rid themselves of them, and soldiers who were mainly misfits in their own regiments, Major Archer was a bitter man.

Gideon went to the provost marshal in the knowledge that thirty Welsh navvies had failed to return to the *Argo*, including Bryn Roughley.

Although Major Archer was in his office, Gideon was kept waiting outside for more than an hour. When he eventually entered the room where the provost marshal was seated behind a desk, he was greeted with a scowl.

'I believe you are the man in charge of the navvies arrested by my soldiers for riotous behaviour.'

'I'm afraid so,' Gideon replied. 'But my understanding is that it was a prize-fight, not a riot.'

'Prize-fighting is itself illegal,' Archer retorted. 'And when my men attempted to apprehend the men involved, others intervened. Their conduct was absolutely outrageous.'

'I would like to offer my apologies on their behalf,' Gideon said, adding, 'There is no excuse for their behaviour, of course, but they are fit men who have been cooped up on a ship for a long time—'

Breaking in on Gideon's apology, Archer snapped, 'Soldiers spend a great deal of time suffering the strictures of shipboard life, but they do not behave in such a manner. I'll see that some of the energy is flogged out of your men tomorrow. That, and a spell in prison, should teach them a lesson that will not be forgotten in a hurry.'

'Flogging? You can't flog my men. They're not soldiers!' Gideon was horrified.

'Malta is under military law, Mister . . . Davey?' Archer found Gideon's response highly satisfying. 'It means they will receive military punishments. It's no more than they deserve and they will remember it for as long as they live. It should also have a salutary effect upon those of their companions who were not apprehended at the time.'

'I am sorry, Major, but I cannot allow these men to be flogged, or to be punished in any way by you. The army has no jurisdiction over them. Indeed, I have a letter which specifically states that the work they are to perform in the Crimea is of such importance that they will *not* be subject to military discipline.'

Gideon's statement took Major Archer aback, but he said, 'I don't know who wrote the letter of which you speak, but I can assure you it will make not an iota of difference to me. *I* am the provost marshal on Malta. *I* make decisions on all matters of law.'

'I have more than one letter in my possession,' Gideon said. 'One from the Duke of Newcastle, Secretary for War, another from the Commander-in-Chief, Lord Hardinge. I also have copies of correspondence between them. Such letters may mean little to you, Major, but I doubt whether the garrison commander will dismiss them lightly. Should he do so I will need to take the matter to the governor. I will also arrange to have a telegraph sent to the Duke of Newcastle, in London, asking him to reply in kind to you personally, ordering you to comply with his instructions. In view of your attitude, I think I should call on the garrison commander now. I am sorry to have taken up your time.'

Aware that he had inadvertently become involved in a matter of far more importance than the unruly behaviour of a number of navvies, Major Archer said hastily, 'Just one moment, Mister

Davey. Do you happen to have the letters of which you speak here with you?'

'I have the letter from the Duke of Newcastle confirming that my men are not to be subject to army discipline. Also a letter from the Prime Minister stressing the importance of the work my navvies are to carry out. In it, he requests the co-operation of all British citizens, wherever they may be. Copies of the correspondence between the War Minister and the Commander-in-Chief are in my cabin on board the *Argo*.'

'May I see those letters you are carrying, please?' Archer's manner now was decidedly deferential.

Reaching inside his coat, Gideon took a stiff manila envelope from a pocket. Extracting two letters, each of which carried an impressive wax seal, he passed them without comment to the provost marshal.

After reading the letters, Major Archer placed them on the desk in front of him, blinking rapidly before staring down at them in deep thought for some time. Eventually shifting his gaze to Gideon, he said, 'May I take these letters and show them to the garrison commander, Mister Davey?'

Gideon shook his head. 'If you wish him to see them I will take them to him myself. They are far too important for me to allow them out of my possession.'

'Of course.' Archer spoke almost humbly now. 'Tell me, why should the War Minister – and, indeed, the Prime Minister – be taking such an interest in your navvies?'

'Due to conditions in the Crimea – conditions I have seen for myself – our soldiers investing Sebastopol are in serious trouble, Major. Very serious trouble. Because of the weather and the difficult terrain it is virtually impossible to get supplies and ammunition to them. A consortium of railway contractors have offered their services to build a railway from Balaklava to the front to relieve the situation. I returned to England to recruit a force of navvies and am now returning with them to carry out the work. I sincerely hope we are not already too late.'

'You have been to the Crimea?'

Gideon nodded. 'I witnessed what everyone in England is

referring to as "the glorious charge of the Light Brigade".' He did not volunteer his opinion that there had been very little glory about the debacle.

'You saw . . . ?' Major Archer rose to his feet and began pacing the length of the room. 'I am an infantryman, Mister Davey, but I would have given my eye teeth to have been there.' Coming to a sudden halt, he scowled at Gideon across the desk. 'Your men behaved in a disgraceful manner, Mister Davey. What is more, three of my men are in hospital as a result of their violence. Many others will have cause to remember them for a very long time.'

'I have already apologised for their conduct, Major. They are rough and rowdy men but, believe me, they will work as no other men in the world can to relieve the suffering of the soldiers around Sebastopol.'

Major Archer nodded his head vigorously. 'Casualties from the Crimea have been passing through Malta in ever increasing numbers – and there have been a disproportionate number of men suffering from cholera and other diseases. I have no wish to do anything to prolong the suffering of our soldiers out there – but I feel very strongly that I cannot allow the conduct of your navvies to be passed over.'

'Your understanding of the situation will be mentioned in my next report to my contractor, Major. No doubt it will be mentioned in London where he has been invited by the Prime Minister to stand for parliament. In the meantime, may I suggest a punishment that will be far more severe for the men involved than it might sound . . .'

Gideon was escorted by two armed soldiers to the communal cell beneath the Valetta castle in which the Welsh navvies were being held, together with a few men from Gideon's own gang who had gone to the aid of Bryn Roughley's men. The place was dark and gloomy and the mood of the navvies matched their surroundings.

The Welshmen, in particular, were far more pleased to see him than at any time since they had sailed from Liverpool.

'Have you come to get us out, Mister Davey?'

'You can take a whole *week's* pay if you persuade 'em to let us out. There are rats in here as big as cats . . . !'

The men crowded round Gideon eagerly, but he said, 'Let you out? I've just been talking to the provost marshal. You've put three of his soldiers in hospital. He wants to flog the lot of you, then give you a long convalescence in his cells. You can't say you weren't warned of what would happen if you got yourselves arrested by the army here.'

The navvies fell silent. The guards had told them they would almost certainly be flogged. It seemed this was Major Archer's favourite punishment for even comparatively minor misdemeanours – and they were aware their conduct put them well outside this category.

'So what is going to happen to us?' Bryn Roughley spoke for the first time, his face grazed and swollen by the beating he had taken at Sailor's hands. 'Will we be sent back to England after we've taken our punishment here?'

'It's what you deserve,' Gideon said. 'But English navvies are no different from Welsh ones when their own are in trouble – and, like it or not, you're *my* men. I've managed to dissuade Major Archer from flogging you . . .'

When the shouts of approval died away, Gideon continued, 'But he's not releasing you tonight. You'll be kept in here and tomorrow morning you'll be escorted back to the ship, immediately before it sets sail for Scutari.'

There was an eruption of sound as the imprisoned navvies expressed noisy relief. Suddenly, it seemed, Gideon was their hero, but Bryn Roughley was not yet completely reassured.

'We're all much obliged to you, Mister Davey, I'll admit that, but how much is it going to cost us this time?'

'It'll cost you your night on the town in Malta, a promise of good behaviour when we dock in Scutari, and a railway built in record time when we reach the Crimea.'

Before the navvies were allowed to leave prison the following morning, Major Archer ensured they were made aware of what would have happened to them had Gideon not succeeded in securing their release. They were taken to the prison courtyard to watch a soldier receive a hundred lashes for insulting an off-duty officer whilst under the influence of drink.

Secured to a specially constructed wooden framework, the soldier broke down less than a quarter of the way through the flogging and began sobbing and begging for mercy. Mercifully, he lapsed into unconsciousness when he had received eighty lashes. However, he would regain consciousness with the certain knowledge that his punishment would be completed when he had recovered sufficiently to take it.

It was a very subdued group of navvies who arrived back on board the *Argo*, escorted by armed soldiers, but the outcome of the fight and its consequences was all that Sailor had predicted. Instead of being two separate gangs, the Welsh and English navvies were now comrades.

Watching them together as the ship headed eastwards, Gideon breathed a sigh of relief. Now that issue had been resolved, he was able to contemplate a reunion with Alice when the ship reached Scutari.

He had not heard from her, but realised that any letters she had written would probably have been chasing him, first to the Crimea, then back to England – and the Crimea had been in a state of chaos when he left.

He wondered whether matters there had improved.

15

I

When Gwen left with Verity to join her husband on the siege lines around Sebastopol, Alice knew she would miss their company, but Nell gave her little time to mope.

In order to help Alice familiarise herself with the items that would be used in the eating-house she was soon to open on the war-torn Crimean peninsula, Nell gave her the task of checking stores from the *Princess* into the warehouse loaned to her by Robert Petrie.

While Alice was carrying out this chore, another large contingent of diseased and exhausted soldiers arrived at the harbour from Sebastopol. The crude front-line hospitals were no longer able to cope with their numbers. Out of a total force of some 23,000 men, more than 10,000 were now on the sick list, half of them requiring hospitalisation. It was the latter group who were being shipped to Scutari.

Alice's task in the warehouse was brought to an abrupt end and responsibility for checking in the stores was passed to Nell's right-hand man, Ali. It came to him with a warning from his employer that if she found she was short of so much as a teaspoon she would have him returned to Scutari.

Alice and Nell devoted their days and much of the night-time to the British soldiers, many of whom were very ill indeed, and any doubts Alice still entertained about Nell were soon dispelled. She was truly what others had represented her to be, a 'doctoress', who put the welfare of needy soldiers above all monetary considerations. Anything in the warehouse that could possibly improve the lot of a disabled soldier was brought out and given freely, with no thought of payment.

For days, the most familiar figures on the crowded Balaklava quayside were the cheroot-smoking Nell Harrup and Alice.

At the height of their activity, Robert Petrie returned from his visits to Lord Raglan's headquarters and the artillery redoubts around Sebastopol. He was able to tell Alice that Gwen had been reunited with her husband, who had been promoted to the non-commissioned rank of bombardier since his arrival in the Crimea.

'It was a very touching reunion, Alice, but I must admit that I was unhappy about leaving her so close to the siege lines. She'll be the only woman with that particular battery and, although there are other women at a nearby compound, conditions up at the front are truly appalling.'

'Gwen will find a way to improve things,' Alice said confidently. 'She's very practical.'

'I'm afraid it will take more than practicality, Alice, it needs a miracle. Things are not likely to get better until we've got the railway to them and the weather improves. That reminds me, there was a telegraphed message for me at Lord Raglan's headquarters. Gideon has left England with the first of his navvies. There will soon be many other ships on the seas between England and Balaklava carrying equipment for us. With any luck, work should be well under way within a month. Unfortunately, I won't be here to see it started. I've been summoned back to London for discussions with the government, but Ranald will give you any help you need. Among the equipment that's already in the harbour is material for building storehouses and huts for the navvies – mainly metal sheets, and the like. Tell Nell she can take whatever she needs to build her

211

eating-house – and a clinic, if that's what she wants. Our soldiers need every comfort that we can give them.'

At that moment one of the soldiers began calling for water and Petrie said, 'I can see you are needed, Alice. I'll say my farewells to you in the morning, before I leave. I have no doubt you and I will meet again in England. Indeed, when all is done in the Crimea you and Gideon must come and stay at my London home. My wife will make you very welcome.'

Late that night, when the last of their patients had embarked on the ships taking them to Scutari, Nell and Alice walked back to Ranald's house, discussing what the contractor had said and what it would mean for them. Nell was full of praise for Petrie.

'He's a good man,' Alice agreed. 'I can remember Gideon saying so. But I wish it was Turnbull who was leaving Balaklava and not Mister Petrie.'

'Has he been pestering you again?' Nell demanded.

'He's always worse when he's had a few drinks,' Alice said, 'which seems to be most evenings just lately. He's under the impression that he only has to promise to give a woman a mention in his London newspaper and she'll fall into his arms full of gratitude. I can't stand the way he edges closer when-ever he's talking to me, either. It makes me want to back away from him the whole time. Still, as long as he does no more than talk I don't suppose I'll come to any harm.'

'If he ever tries to go any further than talking, you must tell me right away, Alice. I've had my share of dealing with men like William Turnbull. I'll put a stop to his nonsense once and for all.'

Two days after Robert Petrie's departure from the Crimea, Ranald MacAllen fulfilled the contractor's promise to take Nell and Alice along the route the railway would take to carry supplies to the troops besieging Sebastopol. With them was the customary escort of cavalrymen, on this occasion a corporal and five dragoons.

Ranald now had his own horse, but such animals were in

desperately short supply. All he could manage to secure for the two women were donkeys.

Never having been taught to ride, Alice was quite happy with this. She felt it would not be so far to fall from the back of a donkey – and she had no doubt she *would* fall.

Nell too was quite content with her mount, declaring that the greatest luxury she had ever known in her young days was the donkey obtained for her by her first husband during Wellington's campaign in Spain.

The dragoons were delighted to be escorting a young woman and they remained in close attendance upon Alice along the way, taking every opportunity to ride close and talk to her.

The pass which led from Balaklava to the valley where the Light Brigade had entered the annals of military history was every bit as bad as it had been when Gideon had travelled the same route. Turkish labourers had been employed in an effort to make a roadway, but their work disappeared beneath the ever-present mud so quickly that it was a waste of time and money, and their attempt ceased.

Once through the pass, Ranald pointed out to Alice where he and Gideon had watched the cavalry charge and rounded up some of the terrified horses to help bring in the wounded. The corporal added that he and the other dragoons had been in the Heavy Brigade which had driven the Russian cavalry from the causeway prior to the Light Brigade's charge, but, although she listened to both men, Alice's thoughts were of Gideon.

She realised that, although Gideon was not a fighting man, he and Ranald had both been very close to the action on that day. The thought that he might have been in any real danger had not occurred to her before.

Since Gideon's departure, Ranald had marked the route of the proposed railway with a line of posts, spaced at intervals across the Crimean countryside. It was these the small party followed and soon they began rising out of the valley and could hear gunfire somewhere ahead of them. One of the dragoons explained that it was a regular duel, fought between the artillerymen of both armies.

'It seems to be a fairly quiet affair today,' he commented. 'I think the Russians are as short of ammunition as we are. Our men get a bonus for each Russian shot they recover, and it's fired back where it came from. No doubt the Russians are doing the same. Some of the cannonballs have probably gone back and forth half a dozen times, at least.'

'It's high time Lord Raglan put a stop to the gunners' little games and sent infantry in to take Sebastopol,' the corporal commented. 'If they'd done it when we first got here, the war would be over by now. All this time-wasting has enabled the Russians to build enough new forts to make it a whole lot harder for the poor infantry when they *are* ordered to take the place.'

'What will the cavalry be doing while the infantry are taking Sebastopol?' Alice asked.

'Making certain none of the Russians from up north come in and attack our infantry from the rear,' said the corporal.

'So the Russians still have an army?' Nell commented.

'Probably as many men as the British and French put together,' the knowledgeable corporal replied. 'And their lines are not too far from here. They extend across the peninsula from just beyond Sebastopol. But their supply problems are even worse than our own. As a result, they don't seem particularly anxious to give us a fight.'

'But they're still out there somewhere,' Nell said thoughtfully. 'I'll need to take that into account when I open my eating-house. I don't want them mounting an unexpected attack and ruining all my efforts.' Turning to Ranald, she asked, 'Where is the railway depot going to be? At Balaklava?'

'No,' came the reply. 'There's a small village halfway between here and the front, and not too far away from Lord Raglan's headquarters. There's water there and plenty of space to keep the rolling stock. Lord Raglan is only interested in the railway as a means of getting ammunition to the troops investing Sebastopol, but if the worst should happen we'll have protection close at hand and the means to evacuate our navvies – and the women, of course.'

214

'Whose idea is it to put the depot close to army headquarters?' Nell asked. 'Robert Petrie's?'

'No, mine,' Ranald admitted. 'But Mister Petrie agrees with me.'

'And so he should,' Nell said approvingly. 'It's good to know there's someone in this war who's thinking ahead and not living on past glories. Where is this village?'

'A short distance ahead. It must have been quite a pretty place before the war began.'

'Right,' Nell said positively. 'That's where I'll build my eating-house – and it'll be called Nell's Nest.'

Ranald was startled. 'But . . . our navvies won't be here for another few weeks. If the Russians *should* attack Lord Raglan won't commit his army to protect an eating-house – not even yours, Nell.'

Nell chuckled. 'When word gets about that Nell Harrup has set up shop I'll have more soldiers around me than Lord Raglan. I doubt whether it will be only Englishmen, either, judging by the number of French and Turkish soldiers we've seen on the way here.'

'They both have larger armies than ours in the Crimea,' Ranald admitted. 'But I'm still concerned about two women setting up an eating-house out here – at least, until the railway's completed.'

'You needn't worry about us,' Nell said positively. 'We'll be all right, you'll see. Nell's Nest will be ready to offer all the comforts of home to your men by the time the railway reaches us.'

Ranald was not convinced and his doubts about Nell's choice of location for her combined restaurant and surgery were not allayed when the corporal said, 'I reckon you'll have more trouble from the locals than from the Russian army, Mother Nell. I don't think they've learned about paying for things they want. You'll need to keep everything nailed down if you don't want it disappearing overnight.'

'Don't worry,' Nell said, 'I've met enough thieves in my time to know how to deal with them. In fact, I'm employing a man

215

who could teach them a thing or two, but let's get on to have a look at this village. I'm eager to choose the site for Nell's Nest and get back to Balaklava to arrange to have the materials Mister Petrie promised me brought up here. Then we can start clearing the warehouse of my things and go into business.'

II

Christmas and the New Year had passed by with only half-hearted attempts to celebrate the occasions, although Nell managed to set up one of her stoves in the dockside warehouse. Here, she cooked a great many Christmas puddings, which she and Alice distributed to the soldiers lying in the small army hospital which had now been set up in Balaklava.

By early January 1855, plans for 'Nell's Nest' were far advanced. Permission had been quickly forthcoming from army headquarters for her to open such an establishment. Indeed, senior officers expressed delight at the prospect of such a diversion from the grim business of waging a difficult and prolonged war against disease, and a dogged and unpredictable foe.

One day, when both women were returning from the site of the eating-house, Nell rode on to the dockside warehouse to discuss something with Ali, while Alice went straight home.

Handing over her donkey to the Turkish servant employed by Ranald, she entered the house, only to come to a sudden, startled halt. An armchair in the sitting-room was occupied by William Turnbull! Reading a newly arrived copy of his London newspaper, he had a half-empty glass of whisky and soda on the table beside him.

Ranald was also in the room, seated at the desk, writing a letter. Looking up, the railway surveyor observed Alice's shocked expression, but before he could say anything Turnbull greeted her jovially.

'Hello, Alice. This is a pleasant surprise for you, is it not?'

'Well . . . it's certainly a surprise,' she admitted warily.

216

'You'll be seeing a lot more of me in the future,' Turnbull said smugly. 'I've moved in. In fact, I've been given the bedroom next to yours!'

Alice's bedroom was reached by a separate staircase to that used by Nell and Ranald, being at the opposite end of the house. Trying very hard not to allow her dismay to show, Alice looked to Ranald for an explanation.

Making an embarrassed gesture of apology, the engineer said, 'It was the least I could do, Alice. I had planned for the railway line to run past William's house, but when I checked my plans again I realised it wasn't feasible. The line needed to go *through* the house. The navvies will be here soon and we need to have as much ready for them as possible, in order to save time, so army engineers are already demolishing the house, and William has moved in here, with us. I meant to tell you a couple of days ago, when the decision was reached, but with so much going on it slipped my mind.'

'It's all right, Ranald. Alice doesn't mind, do you, my dear? We're old friends from Scutari . . . but where are my manners? You must have a drink to celebrate the occasion. I've brought a plentiful supply with me . . . what will you have?'

Declining his offer, Alice pleaded the need to clean up after a day spent in the mud of the Crimean countryside. Making her way to her bedroom she felt dismay at the thought of having Turnbull living in the same house. His clumsy advances had embarrassed her on more than one occasion during their time in Scutari and she had tried to maintain a safe distance between them since coming to Balaklava.

The situation would improve again once she and Nell had moved out of Balaklava – or when Gideon arrived – but both events might still be some time away. Meanwhile, she and Turnbull would be living beneath the same roof. It was not a comfortable thought.

Alice's fears about Turnbull's behaviour were fully justified only a few nights later. Ranald was spending a few days at Lord Raglan's headquarters, having detailed discussions with military engineers and mapmakers, while Nell had been invited to a party in honour of one of the army doctors with whom she was friendly, who was returning to England on promotion.

Turnbull had put in an early appearance at the same party and remained there for some time, even though he was ignored by a number of the medical staff, many of whom had been strongly criticised in his newspaper articles.

After consuming a considerable amount of alcohol, the correspondent left the party and made his way to a hostelry frequented by the naval men who called at Balaklava. He remained there for a while before heading, unsteadily, for the house rented by Ranald.

Not used to being in the house by herself, Alice had taken the unusual step of bolting her bedroom door from the inside. She was thankful for the precaution when she heard Turnbull come in and make his way upstairs. It was apparent from his noisy progress that he had been drinking heavily.

Mumbling to himself, the newspaperman paused on the landing. Then, to Alice's dismay, he banged heavily on the door of her room.

'Alice . . . ? Alice, are you in there, girl? I've brought some drink home for you . . . gin. I thought you'd like to have a drink with me. Shall I bring it in?'

The slurring of his words confirmed Alice's assessment of his intoxicated state and she made no reply.

'Alice . . . are you there?'

The latch was raised but the bolt prevented the door from opening.

'Alice . . . I know you're in there. Come on, girl, open up. You must learn to relax a little. You know what they say about all work and no play. Come on now, have a drink with me, and enjoy yourself . . .'

The latch rattled again and Turnbull put his weight against the door, but it was made of stout wood and the bolt held. Nevertheless, Alice's heart was beating at an alarming speed. There was no one else in the house and William Turnbull was far too drunk to be reasoned with.

Then she heard something fall to the floor on the landing outside her room and the drunken correspondent began grumbling unintelligibly. It sounded as though he picked up the object only to drop it again, and now he began to swear. But then she heard him move away. Moments later there was the sound of the door to his own room being opened, then slammed noisily shut.

Alice realised she had been holding her breath. When she released it, she began shaking and she did not relax until she heard Nell come home.

William Turnbull had not emerged from his room by the time the two women were taking breakfast and Alice told Nell what had occurred when he had returned to the house the previous evening.

Frowning, Nell asked, 'Are you quite sure he was trying to get into your room, Alice? Could it not have been that he was drunk and fell against your door?'

Alice shook her head vigorously. 'No. He tried the latch a couple of times and told me he'd brought some gin back to the house for me.'

'Did he now! Well, we can't allow Turnbull to get away with this, Alice, especially as we won't be moving out until Nell's Nest is ready. We'll need to think of some way of putting an end to Mister Turnbull's attentions.'

'Perhaps what happened last night won't happen again,' Alice said hopefully. 'The party he'd been to could have been for someone special.'

'I doubt it,' Nell replied. 'There are many such parties. It's always the same in wartime whenever the opportunity arises – and a promotion's as good an excuse as any other, though I can't for the life of me see why it should be. Any officer who shows promise is promoted and most likely gets sent home to waste his time counting rations, or whatever they do in offices there. Meanwhile, others with no experience and little talent are sent out to the front, eager to lead troops into action so they too can get promoted and sent home. At least, that's what my last husband used to say and I've seen nothing in my time with the army to convince me he was entirely wrong. But to get back to Turnbull . . . he'll no doubt be invited by senior officers to all their parties, in the hope that he'll give them a mention in his reports to his London newspaper. In fact, I've heard there's a big party taking place tomorrow night, for someone who's been promoted to brigadier.'

'Oh! Perhaps I can find somewhere else to stay,' Alice said unhappily.

'I don't think there'll be any need for that, dearie. I've got an idea. I think we might be able to cure William Turnbull's amorous designs on you, once and for all . . .'

III

The brigadier's farewell party was a lavish affair. Even in his usual drunken state, William Turnbull could look back upon it with considerable satisfaction. He had been kept liberally supplied with drink – good champagne and brandy, obtained from the French army – and officers had listened respectfully when he propounded his views on how the war should be conducted. What was more, he knew there would be many such parties to enjoy in the coming weeks.

Such were the thoughts of the correspondent as he left the officers' mess where the party was being held. Once outside, the cold air hit him and he had to concentrate even more than usual in order to maintain a zig-zag course in the general direction of Ranald MacAllen's house.

When he arrived, he needed to aim himself at the front door and, once inside, he viewed the stairs to his bedroom with some trepidation.

Three times he attempted to climb them, only to overbalance on the second or third stair and fall back to the hallway. He finally succeeded in reaching the landing by tackling the stairs as he would a steep mountain slope, mainly on hands and knees.

Pleased with his success, Turnbull rose to his feet. Clutching at the banisters for support, he tried to will both eyes to focus upon the same object. It was not easy. There was no light up here, the only illumination coming from a dim lamp in the hallway downstairs. However, it was sufficient for him to see that Alice's door was slightly ajar.

Checking his pockets clumsily, he located the half-full bottle of brandy that, as had become customary with him, he had appropriated from the officers' mess.

Swaying forward, he knocked on the open door. 'Alice, are you there?'

'I thought you might call. Come in.'

Delighted with the unexpected response, Turnbull pushed the door fully open and stumbled inside the darkened room.

'I left the party specially early, Alice. Brought a drink for you.'

'That's very kind of you, but close the door behind you in case someone comes in downstairs. I don't want my reputation ruined.'

'But . . . it's dark, Alice . . . I can't see you.'

'Why do you need to see? The bed's right in front of you . . . That's it. Look, you get undressed and into bed and I'll pour drinks for us. Give me the bottle.'

There was more than a hint of nervousness in Alice's voice, but Turnbull was far too drunk to notice. Fortunately, her eyes were accustomed to the darkness of the room and, taking the bottle from his hand, she easily evaded his clumsy attempt to embrace her.

'Aren't you going to give me a kiss, Alice?' Turnbull pleaded.

'There's plenty of time for that. You undress and get into bed before I change my mind.'

Turnbull undressed clumsily, falling about as he did so. The darkness did not help his already disoriented state, but eventually, wearing only knee-length cotton drawers, he climbed on to the bed and pulled the bedclothes up around him.

'I knew we'd get together one day, Alice. I knew it . . . come to bed now.'

'I'm going to enjoy a drink first. You too. Here . . .'

221

Holding it at arm's length, Alice handed him a tumbler half filled with brandy.

Turnbull took the glass from her clumsily, grumbling when she evaded his other hand.

'Go on, drink it up – I'm drinking mine,' she lied.

There was silence for a minute or two before Turnbull slurred, 'My drink's all gone . . . haven't you finished yours yet?'

'Nearly. Be patient. A girl doesn't like to be rushed on such an occasion. Surely you're experienced enough to know that?'

'Of course . . . you take as long as you like, Alice . . . I won't disappoint you . . .'

It seemed William Turnbull was struggling to find more words, but Alice could make out his body swaying backwards and forwards as he sat up in the bed. Suddenly, the glass slipped from his grasp and dropped from the bed to the floor. At the same time, Turnbull fell backwards and lay still.

It all happened so suddenly that it frightened her. However, after reassuring herself that the correspondent was still breathing, Alice retrieved the glass from the floor and, placing it upon a table, slipped quietly from the room.

When William Turnbull woke the next morning it felt as though his head was being painfully crushed by a giant hand. Even the thought of opening his eyes caused him to wince. His mouth seemed devoid of all moisture and was filled with a tongue that felt like a piece of old leather, twice its normal size.

It was some minutes before his mind could put a series of disjointed images into any coherent sequence. It would have taken longer had a movement in the bed alongside him not speeded up the process.

'Alice . . . !' He turned over quickly – and wished he had been more circumspect as the room gyrated wildly about him.

As he paused, wincing with sudden pain, a gruff voice asked, 'Why are you calling for Alice when you've got me . . . William?'

'Nell! Nell Harrup. What are you doing here?' Turnbull sat bolt upright, ignoring the pain and the nausea as the room spun about him once more.

'Well! What sort of question is that – after last night?'

'After last night . . . ?' Turnbull tried to marshal his scrambled thoughts. 'What are you talking about? What's happened to Alice?'

'Why do you keep talking about Alice? If you must know, a party of Florence Nightingale's nurses arrived in Balaklava yesterday and a friend of Alice was among them. Alice helped her settle in to her new quarters and stayed with her last night. The wind rattling the windows in my room has kept me awake for the last couple of nights, so I thought I'd come to this side of the house to get a good night's sleep – but I'd reckoned without you, William. Not that I'm complaining, mind you! I thought I was past all that sort of nonsense, but I'm happy to know I'm not! In fact you've whetted my appetite all over again . . . come here, lover . . . !'

Moments later William Turnbull stumbled out of Alice's room into his own, clutching his clothing to him and pursued by Nell's raucous laughter.

Nell was still chuckling when she mounted the stairs to her own bedroom. Entering, she found Alice sitting on the bed, fully clothed and looking worried. 'Is everything all right, Nell?' she asked anxiously.

'Of course. I haven't had such a laugh for years. I doubt whether Mister Turnbull will bother you again. I convinced him it was me who invited him into your room last night and let him think he'd had his way with me.'

'And . . . did he? I mean . . . did anything happen?'

Getting dressed as she spoke, Nell replied scornfully, 'If I ever did that again it would be with a *real* man, as all my husbands have been. It wouldn't be with the likes of William Turnbull. Still, I'll give him a knowing smile whenever we meet and we'll see how long he can take sharing the same house with us. But what's the time, Alice? That stuff I gave you to put into Turnbull's drink must have been stronger than I thought. I seemed to be lying there for ages, waiting for him to wake up.'

'It must be half past nine, at least.'

'As late as that!' Nell exclaimed. 'Come on, girl, we've had

our fun, now it's time to get down to some work. I was telling
Turnbull the truth when I told him about some of Florence
Nightingale's nurses arriving in Balaklava. So now we can go
up to Nell's Nest with a clear conscience – and there are more
men to be saved there than here.'

Nell never found it necessary to give William Turnbull one of
her 'knowing smiles'. When they came back from Nell's Nest
that evening Ranald MacAllen had returned from Lord Raglan's
headquarters. He told them Turnbull had taken passage to
Scutari that same afternoon – called there on 'urgent business'.

It was a much happier Alice who went to bed in her own
room that night, especially as Ranald had received a telegraph
message confirming that all the navvies had now set sail from
England, together with the bulk of the materials needed to build
the Crimean railway. The message also said that Gideon's ship
was likely to arrive at Scutari at any time.

16

I

When the *Argo* dropped anchor off Scutari on a cold but crisp January morning, Gideon was already on deck gesturing for one of the many Turkish boatmen touting for business to come alongside and take him ashore.

The *Argo* had made good time from Malta and, as Sailor had predicted, the fight that had taken place between himself and Bryn Roughley had cleared the air between the two gangs of navvies and the two combatants had become good friends.

Gideon now felt confident he had the nucleus of a workforce on which he could rely, one equal to any other in the world. He need have no feelings of guilt for devoting some of his time to personal matters. To Alice.

His eagerness to go ashore was witnessed by Sailor and Bryn Roughley as they stood together smoking their pipes in the lee of a small deckhouse.

'The boss seems unusually anxious to get off the ship,' Roughley commented. 'Do you think it's because he wants to get away from us for a while? And here's me thinking we were all finally one big, happy family.'

Sailor smiled. 'I don't think Gideon has work on his mind

today. He's gone ashore to meet up with a girl – a very nice Cornish girl who came out here with Florence Nightingale's nurses. She's probably going to be Mrs Gideon Davey one day, and they'll make a handsome couple.'

'You think a lot of Davey, don't you? Any special reason?'

'More than one. He's a good ganger, probably better than you and I will ever be. He's honest and fair, and when the chips are down he's tough – and I mean *really* tough. I wouldn't rate the chances of any man if he took Gideon on in a fight over something that really mattered to him. Having said that, he'll fight as hard for you, me, or one of his gang, as he will for himself.'

'I think he proved that in a quiet way when we got ourselves arrested in Malta,' Roughley said ruefully. 'My gang certainly think so.'

'Your gang are right,' Sailor agreed. 'But the only time I've ever seen Gideon set out to really destroy someone was when he fought a Cornish wrestling champion who'd made life difficult for the girl he's going ashore to see now. He gave him the beating of his life.'

Knocking out his pipe over the side of the ship, Roughley commented, 'From all I've heard about conditions in the Crimea he'll need to be as good as you say he is if he's to build a railway there. But there's one thing in his favour. He's got the two best gangs of navvies in the business to help him.'

Gideon made his way up the hill towards the Scutari military hospital, beset with conflicting emotions. Uppermost was excitement at the thought that he might be only minutes away from being with Alice again and trying to guess what her reaction would be. During their long parting the brief happy times they had shared had grown in his mind, especially the outing he had arranged as a birthday treat for her.

He believed that had there not been the unfortunate incident involving the paint-daubed door on their return to the rectory, the day might have ended with a firm understanding between them.

As it was, Gideon had a nagging fear that she might have

found someone else since their last meeting, especially as life had changed so much for her during that time. The Reverend Harold Brimble, of Bodmin, had assured him that Alice had been most concerned lest her move from the Treleggan rectory should result in their losing touch with each other, but that had been a number of months ago. Gideon had received no mail from Alice – and could not even be certain the letters he had written had reached her.

He was aware that she was no longer a young servant-girl with little knowledge of life beyond the confines of an isolated Cornish rectory. She had now travelled farther from England than most people did in a lifetime, during the course of which she must have met scores of soldiers who were far from home and starved of female company.

Gideon was not vain enough to believe he had more to offer her than many of those she had encountered since their last meeting.

Overawed, as Alice had been, by the sheer size of the hospital building, Gideon halted at the main entrance, at a loss where to begin his enquiries.

In the midst of his indecision, a group of nurses left the building together, chattering in English. Hurrying after them, Gideon spoke to one who had an air of authority about her.

'Excuse me, I'm looking for someone who I believe is a nurse here. Her name's Alice . . . Alice Rowe. I wonder if you could tell me where I might find her?'

The Honourable Josephine Wotton looked at Gideon distastefully. 'We have no Nurse Rowe here. There was an Alice with us for a while, but she was employed as an orderly, and she left some time ago.'

'Alice has left!' Gideon was aghast at Josephine Wotton's words. 'Where has she gone?'

Tight-lipped, the aristocratic nurse said, 'I suggest you speak to one of the Sisters of Mercy. Alice was in their employ . . . there's Sister Frances coming from the hospital now.' And with that, Josephine Wotton hurried after the other nurses.

Gideon turned to see a black-garbed nun coming down the steps of the hospital and he hurried to intercept her. 'Excuse me, Sister Frances, I wonder if you can help me?'

Gideon thought the stern-featured nun somewhat daunting, but when she looked directly at him he thought he detected something gentle in her eyes that contradicted his first impression of her.

'If it's something that lies within my power, I will be happy to do so.'

'I'm looking for Alice Rowe. I asked one of the nurses who left the hospital just now. She said you might know her whereabouts.'

There was a guarded expression on the face of Sister Frances as she asked, 'Why do you wish to find Alice? Are you one of the men she has nursed?'

'No, Sister . . . I'm sorry, I should have introduced myself properly. My name is Gideon Davey. I knew Alice in Cornwall but had to leave unexpectedly to come to the Crimea for Mister Petrie, a railway contractor. I returned to England expecting to find Alice there – but was told by a Reverend Brimble that Alice was out here! We had somehow managed to miss each other. It appears that it might have happened again!'

Sister Frances's pleasure was apparent as she rested a hand on Gideon's arm. 'So you are the young man Alice spoke so much about. I am delighted to meet you – and Alice will be too. Unfortunately, she is no longer here, at Scutari.'

'She's not . . . ? She hasn't gone back to England?'

'No, it's not quite as serious as that – although I would be happier were it so. She is in the Crimea, in the company of Nell Harrup, a woman who calls herself a 'doctoress' and who, I hasten to add, is highly thought of by those she has treated, even though no place could be found for her here, with Miss Nightingale's nurses. I believe Mrs Harrup intends opening an eating-house to offer comfort as well as medical care to our soldiers in the Crimea. She persuaded Alice to accompany her there.'

Gideon felt there was more than a hint of disapproval in Sister

Frances's voice when she was talking of Nell Harrup. 'I and my men will be leaving for the Crimea tomorrow. Do you know exactly where I might find her there?'

'No . . . but I was talking only yesterday to a Mister Turnbull, a newspaper correspondent. He has recently returned from the Crimea and Alice's name came up in our conversation. He may be able to help you. I think you will probably find him at the officers' club, down the road from here.'

After promising to pass on best wishes to Alice from Sister Frances, Gideon went in search of William Turnbull. He was concerned about Alice, not only because she was in the Crimea, but because he sensed Sister Frances's disapproval of Nell Harrup, even though the nun had insisted that Nell was 'highly thought of'.

II

On his way to the club where he expected to find William Turnbull, Gideon thought of what Sister Frances had said to him. Her reservations about Nell Harrup were disturbing, but he hoped William Turnbull might be able to put his mind at ease.

He was doomed to disappointment.

William Turnbull was drinking on his own and, although it was not yet evening, he was quite obviously inebriated. He explained his condition by telling Gideon he was celebrating his return from the Crimean battle front by partaking of 'an after lunch drink'. He failed to add that his celebrations had already lasted for more than a week.

Greeting Gideon as though they were life-long friends, he stood up unsteadily to shake his hand with great gusto before waving him to a vacant seat at his small table and pouring him a very large brandy from the bottle standing there.

Seated once more, Turnbull beamed at Gideon over the rim of his goblet. When he had taken a generous swig of his drink, he leaned back in his chair and said, 'So you've come back to the Crimea to build your railway? I wish you luck, young man – but you'll need all the luck you can find. The army have been trying to make some sort of road to the front, but all their efforts are lost beneath the Crimean mud. I fear your railway will be no different.'

'You underestimate my navvies,' Gideon said, taking a cautious drink from his glass, 'but right at this moment I have other things on my mind. I expected to find Alice Rowe here, at Scutari. When I went to the hospital just now, I spoke to a Sister Frances. She says Alice is in the Crimea and suggested you might know exactly where I can find her.'

'Ah! The delectable Alice.' Turnbull leered at Gideon in a manner that made him feel extremely uncomfortable. 'She and I became friends . . . extremely *good* friends. As we were sharing the same house it would have been surprising had it not been so. I used to take a drink back to the house for her on most nights, after I had been out. She appreciated it.'

Gideon was taken aback by Turnbull's statement. When he had taken Alice to Falmouth she had declined strong drink, even when they were having a meal together. Aware that much of her time since then had been spent working with the Devonport Sisters of Mercy, he doubted whether she would have developed a taste for alcohol in their company.

'Is Alice still in Balaklava?' he asked.

'She was when I left the Crimea,' Turnbull replied, 'but she and Nell Harrup were planning to move to a village halfway between Balaklava and Sebastopol – to provide "comforts" for the troops.'

The emphasis Turnbull put on the word was not lost on Gideon. 'What exactly do you mean by providing comforts?' he asked sharply.

Turnbull shrugged. 'I suppose the word conveys a different meaning to whoever happens to be using it. For Alice, it no doubt conjures up visions of tea-houses and parties on a vicarage lawn. I think Nell Harrup might have a very different interpretation.'

Trying to hide his concern, Gideon frowned. 'What do you know of this Nell Harrup?'

'Ah! Our Nell is very much a woman of the world. She's been a camp-follower for all of her life, and has had at least three soldier husbands – although whether she received the blessing of the Church for any of them is another matter entirely!'

Gideon thought that if Nell Harrup was as William Turnbull had described her, she was hardly the type of woman with whom Alice would strike up a friendship.

When he voiced his thoughts to Turnbull, the correspondent replied, 'Oh, I'm not saying Nell is entirely bad. There are soldiers, officers and men, who have more faith in Nell than in their own doctors – not that anyone should find that particularly surprising – but her personal life leaves much to be desired.'

Trying not to think too much of what the correspondent had said, Gideon asked, 'Where in Balaklava was Alice living, when you left?'

'In the house occupied by your colleague, Ranald MacAllen. Nell's living there, too.'

Turnbull's reply took Gideon by surprise. Not because Alice was staying there. Ranald was aware how much Gideon thought of Alice. He would have done everything in his power to help her once he knew she was in Balaklava. But if this Nell Harrup was all that Turnbull claimed her to be, then Gideon could not understand why Ranald had accepted her in his house.

It was equally surprising that Turnbull should also have been invited to lodge in the house. Gideon and Ranald had gone out of their way to stay clear of the correspondent during their early days together in the Crimea. They had both agreed William Turnbull was a troublemaker and generally disliked by all who came into contact with him.

Remembering this made Gideon sceptical about the veracity of Turnbull's stories of Alice, but he wished he had been able to learn more about Alice's colourful companion.

231

III

Much to Gideon's relief, the *Argo*'s brief stay at Scutari passed without any serious trouble. Three of the navvies – two Welshmen and a Cornishman – were arrested for drunkenness, but Gideon's intervention was not required in order to secure their release. After spending the night in Turkish police cells they were sent back on board before the ship sailed.

It was a very satisfactory state of affairs, but when Sailor found Gideon standing on deck by himself, soon after the *Argo* left Scutari, he sensed he was not happy.

Standing beside him, packing tobacco into his pipe, Sailor said, 'Is the thought of building this railway worrying you, Gideon? There's no reason why it should. All the navvies are behind you and determined to show the army what they can do.'

'I have complete faith in the men, Sailor. Besides, Robert Petrie has been out here and much of the preparatory work will have been carried out before we arrive. No, I'm bothered by a more personal matter.'

He told his companion what Turnbull had said about Alice and Nell Harrup, adding, 'I find it difficult to believe that Alice would drink with Turnbull late at night, but I'm concerned about the influence Nell Harrup might have had on her. I could tell that the sister I met at the hospital was uneasy about it too.'

'Well, I've only met Alice the once,' Sailor admitted, 'but from what I saw and what you've said about her I wouldn't have thought she'd let herself be influenced by the wrong sort of people. Billy Stanbeare couldn't do it and he must have tried for years. As for Nell Harrup . . . I wouldn't expect someone like this sister at Scutari to even begin to understand her.'

Gideon looked at Sailor sharply. 'Do you know Nell Harrup?'

'I know of her,' Sailor said. 'She was with the Victoria barracks garrison in Hong Kong when I was out that way very many years ago. She was reckoned to be a tough lady in every

way – something of a rough diamond – but she was known as
the Queen of Victoria. The soldiers thought the world of her.
She worked day and night when they had a serious epidemic
of something or other. If Nell is out here, then it's to do what
she can for the soldiers and Alice will be helping her.'

Gideon remembered that Sister Frances had admitted that
Nell Harrup was highly thought of by the army. He decided he
would reserve judgement upon her until he found Alice. He
would then make up his own mind about the woman with
whom she was working.

As the *Argo* approached Balaklava, it ran into a fierce storm.
Most of the navvies had gained their 'sea legs' on the long
voyage from Liverpool and were not ill, but the ship's captain
felt it advisable to order them to remain in their quarters and
not venture on deck.

Gideon remained below with them until he was called on
deck by the captain as they approached the inlet that formed
the anchorage outside the small Balaklava harbour.

The captain explained the summons by saying, 'This is the
first time I've sailed into Balaklava. The approach to the harbour
looks quite tricky and as my chart is copied from an old Russian
one I don't entirely trust it. I'm also concerned at passing so
many ships who seem to prefer riding out the storm to
remaining in the anchorage.'

'There was a disastrous storm here a couple of months ago,'
Gideon explained. 'It was the worst the area has ever known
and wrecked a great many ships when they dragged their
anchors. It looks as though no one is taking any chances this
time.'

'Shall I ride it out, or do you think we can make the harbour
safely with a wind like this behind us?' The *Argo*'s captain was
clearly concerned.

'You've taken in most of your sail,' Gideon said, 'so with
virtually no shipping in the anchorage we should be all right.
The harbour itself is more sheltered and there is a steam tug
to help us to go alongside. Talking of steamships, there's a

steam warship coming out of the anchorage and heading straight for us.'

The warship was one of the Navy's latest iron-clad vessels, of a design quite unlike any ship Gideon had ever seen before. With smoke belching forth from a single tall funnel, it sliced through the rough sea with comparative ease, heading towards the *Argo*.

The captain displayed little concern. 'I'll have the bell sounded, but it's an accepted rule of the sea that more manoeuvrable steam-powered vessels give way to sailing ships.'

At first, it seemed the captain's confidence was justified. The warship altered course and was about to pass close on the starboard side of the *Argo* when suddenly it veered violently to its right, heeling over in a tight circle.

'Its rudder's gone!' the *Argo* coxswain cried. Without waiting for an order from the captain, he heaved on the wheel in an attempt to avoid a collision. Gideon leaped forward to help . . . but it was too late. All they could hope for now was that the *Argo* could be turned sufficiently to take no more than a glancing blow from the errant warship.

Their efforts achieved only partial success. The force of the collision rolled the *Argo* almost on to its side. Had the captain not grabbed the wheel at the last minute he would have slid into the sea from the sloping deck.

The noise of splintering timbers was horrendous but the impact would have been far worse had the captain of the warship not realised at the last minute there was going to be a collision and ordered the engines to be thrown into reverse.

The *Argo* righted itself quickly, but the next few minutes were chaotic and confused. As men clambered from the lower decks, wondering what had occurred, one of the crew shouted that the *Argo* was taking in water – fast!

'Shall I give the order to abandon ship?' the coxswain demanded of the captain.

'No!' Gideon spoke before the captain could reply. 'The *Argo* is carrying all my navvies' tools, as well as a lot of vital equipment. Lose that and the railway will be set back by a couple of months, at least. The ship has to be saved.'

'I've got a sinking ship and I need to think of my crew—'

'Crowd on all the sail and run the *Argo* aground,' Gideon interrupted the captain. 'There's a gently sloping beach just to the right of the harbour. It's where the villagers keep their fishing boats. It won't do *them* a lot of good, but you'll save your ship – and my equipment.'

The *Argo*'s captain realised immediately that Gideon's idea might work if he could build up enough speed – and if the *Argo* remained afloat long enough to make it possible. 'Shake out every sail we have,' he shouted to his officers. To Gideon he said, 'If this beach isn't too steep, with this wind behind us it should take us far enough out of the water to save the *Argo* – and your stores. We'll at least have tried. Get your navvies down below and try to stop water coming in. Use everything that comes to hand. Hammocks, mattresses. Anything that'll block the hole. I'll get the pumps going and give your men a warning when we're about to go ashore.'

The next minutes were hectic ones as the *Argo* gathered speed along the narrow anchorage that led to the harbour. The irony of the situation was that had it not been for the storm the anchorage would have been crowded and such a desperate attempt to save the vessel would not have been possible.

The beaching of the *Argo* was something of an anti-climax. Warnings were shouted to the navvies moments before the ship struck and they either dropped to the deck or clung to anything that was available to prevent themselves being flung against a bulkhead and suffering injury. Such precautions proved unnecessary. The shore here was a gently sloping pebble beach and the *Argo* struck with sufficient speed to carry more than half its length clear of the water. It came to rest leaning rather alarmingly to one side, but that could be righted with a little effort. Everyone on board was aware they had had a very lucky escape.

The captain was jubilant. Although badly holed, his ship was capable of being repaired. His only doubts were of getting the vessel back in the water. The Black Sea was not tidal and it would not be possible to make use of a particularly high tide to refloat the *Argo* when repairs had been carried out.

'That will be no problem,' Gideon said, when the captain voiced his fears. 'When the ship is ready to put to sea again, I'll bring my navvies down here to dig a channel deep enough to float it and the steam tug will pull you out into the anchorage. In the meantime we'll see about shoring up the ship. With your permission, I'd like my navvies to remain on board while repairs are carried out. At least, until they have a camp built onshore.'

The captain agreed immediately. It would provide additional security for the *Argo* and he knew it would probably be needed.

There had been a number of small fishing boats pulled up on the shore, and when the *Argo* arrived in such a dramatic manner the owners began clamouring around the ship bewailing the loss of their livelihood, sensing a substantial profit by way of compensation.

Many of the inhabitants of the small coastal town also came to view the grounded *Argo*, as did several of the off-duty soldiers based in and around Balaklava. The British admiral in charge of the harbour put in an appearance too. After praising the captain's presence of mind in saving his ship, he promised a full investigation into the part played in the collision by the warship.

IV

Ranald MacAllen came on board the *Argo* soon after the admiral and, after marvelling at the ship's lucky escape, greeted Gideon warmly, adding, 'Lord Raglan is going to be delighted to have you and the men here. Things are going from bad to worse up at the front. The weather has been atrocious. He's lost so many men through sickness that he's had to swallow his pride and ask the French army for help. They're carrying ammunition up to the front for the siege guns and bringing back incapacitated soldiers when they return to Balaklava. As a veteran of the war against Napoleon, that sticks in his craw more than the lack of progress of the war itself. Fortunately for everyone, the weather seems to be affecting the Russians almost as seriously as ourselves. If they were in a position to attack right now they would drive us out of the Crimea. We've got to have this railway built in record time, Gideon.'

The seriousness of the situation was not lost upon Gideon, but for the moment he had other matters on his mind. He asked Ranald for news of Alice.

'It's a pity you didn't arrive in Balaklava a couple of days ago,' Ranald said. 'She and Nell Harrup have moved into Nell's eating-house – "Nell's Nest". It's in an inland village, not very far from Lord Raglan's headquarters.'

'I was told at Scutari that something of the sort was planned.' Gideon felt frustration and disappointment that his reunion with Alice was to be postponed yet again. 'But I haven't been given any details. In fact, I've heard nothing from Alice since I left England for the first time.' Bitterly, he added, 'I sometimes wonder whether she still remembers me.'

'That's one of the most ridiculous things I've ever heard!' Ranald said incredulously. 'A day hasn't passed when she hasn't spoken of you. As for hearing from her . . . I've sent off at least three letters on her behalf, all written to you at a time when she and Nell were working all hours tending wounded soldiers on the quayside, right here in Balaklava.'

'They must still be chasing me back and forth between here and England,' said Gideon, chastened by Ranald's words. 'But this Nell's name keeps cropping up in any conversation I have about Alice. What's the true story about her, do you know?'

'I've heard enough stories about her to fill a book,' Ranald replied. 'Many are little more than exaggerated rumour, but I've seen enough of her to realise she's a remarkable woman. She's travelled the world as an army wife, and on the way learned more about medicine than most army doctors. The soldiers she's tended – both officers and men – think the world of her. The reason she's opening her eating-house so close to the front is so that she can provide a little comfort for the men and be on hand to treat them when they need her. There's no doubt that any profit she makes from Nell's Nest will go towards providing medicines and other things for whoever needs them. She'll never be a rich woman, Gideon, but, by God, she'll never be forgotten.'

'You have a higher opinion of her than others I've spoken to, Ranald. I've heard it suggested that her morals are particularly suspect.'

'Then you've been talking to the wrong people,' Ranald said bluntly. 'She's unconventional, certainly, and she doesn't give a damn what others think of her, which is why she hasn't been accepted by Miss Nightingale and her nurses, but she's a very special woman. You can take my word for that. We were living in the same house all the time she was in Balaklava.'

'You must have had quite a houseful,' Gideon commented. 'I believe both Alice and William Turnbull were living there too. I met Turnbull in Scutari. He claimed he and Alice became regular drinking partners while they were in the house.'

Looking quizzically at Gideon, Ranald said, 'I hope you didn't believe a word of it. Turnbull made quite a nuisance of himself to Alice – until Nell put a stop to it in her own inimitable way.'

Ranald went on to tell Gideon what had happened on William Turnbull's last night in Balaklava. Gideon was both amused and relieved. He told himself he had never really believed Turnbull's boasts about his close relationship with Alice, but there had always been a nagging awareness that he and Alice had really spent very little time together before he had been forced to leave Cornwall. There had been no time to learn everything about her.

'Talking of Nell . . . I saw her at the harbour only a few minutes ago. She's arrived with some of the men brought from the front late this morning. There's been a fairly brisk skirmish with the Russians and our soldiers have suffered a number of casualties. The army are clearing the forward hospitals of those men not likely to recover quickly and shipping them off to Scutari. Go and ask her about Alice. I'll take care of things here for a while.'

'Thanks, Ranald. If there's anything you need to know, ask Sailor. He'll be able to give you the answers.'

* * *

Gideon's first sighting of Nell Harrup seemed to confirm Ranald's opinion of her and give the lie to what William Turnbull had said about the 'doctoress'.

Nell was crouching beside a young soldier, holding his hand and occasionally stroking his hair, at the same time talking to him in a low voice. Gideon watched until she slowly released her hand and gently drew the blanket up to his chin before standing up.

She was immediately called by another soldier, and after giving him a drink she filled a pipe taken from his pocket with tobacco that she carried in a large pouch, lighting the pipe before handing it back to the man whose hand was shaking so much he was unable to light it for himself.

Then she hurried off to a distressed soldier who looked to be no more than a boy. After speaking briefly to him, she removed the bandage that was wound about his head and changed the dressing covering a bullet wound that had carved a deep furrow across his forehead.

Gideon made his way towards her, but by the time he reached her she was holding a stretcher-borne soldier in a sitting position, in order to give him water.

She did not look up when Gideon approached her. After standing awkwardly beside her for a while, he said, 'Hello. I believe you're Nell Harrup.'

'That's right,' she said, still not looking up at him. 'I've noticed you watching me at work among the men. Who are you, someone from the army medical department who wants to stop me from carrying out the work your doctors ought to be doing? Or are you perhaps a newspaper correspondent, trying to discredit either me or the army doctors?'

'Neither,' Gideon replied easily. 'I'm a railway man who's trying to find Alice Rowe.'

When Nell finally glanced up at him, Gideon looked into the face of a very tired lady. Despite this, he felt she was giving him a thorough appraisal.

'Why are you looking for Alice? Do you have a message from someone for her.'

'Not exactly,' Gideon replied. 'I want to find out whether any of the letters I've sent have actually reached her.'

He had Nell's full attention now, but before speaking to him again she gently laid the soldier back on his stretcher. Straightening up, she said, 'You must be Gideon – the man she's been talking about for as long as I've known her.'

Gideon smiled. 'I'm certainly Gideon – and I'm very relieved to know she hasn't forgotten me.'

'Forgotten you?' Nell gave him a quizzical look. 'There are times when I despair of her thinking of anyone or anything else. How long have you been in the Crimea?'

'Little more than an hour,' Gideon replied. 'After a collision as we entered the anchorage we had a dramatic arrival, being forced to beach our ship among the fishing boats pulled up alongside the harbour.'

'I heard some of the army men talking about it,' Nell said. 'Have you come with your men, ready to build the railway the army have been waiting for?'

Gideon nodded. 'The first of my navvies are here with me. We'll have started work by the time the others arrive.'

'It won't be a moment before time,' Nell said. 'I know its main function will be to carry ammunition up to the siege lines around Sebastopol, but there'll be room in the wagons to bring badly wounded men back here to Balaklava. It will save a great many lives.'

'I have no doubt at all that you'll be able to persuade Lord Raglan to do whatever's needed,' Gideon said, 'but my job is to build the railway. It will be up to him to decide what is done with it – but all that will begin tomorrow. Right now I want to know how quickly I can meet up with Alice. It seems that every time I arrive at a place where I expect to find her, she's moved on somewhere else.'

'I doubt if you'll have any more problems with that,' Nell said. 'If I'm not mistaken the next batch of casualties are just arriving from the front – brought here by the French and their mules. Alice should be with them.'

17

I

The evacuated British soldiers were brought from the front to Balaklava, two at a time, on mules fitted with specially constructed saddles, the animals led by French colonial troops recruited in North Africa. It was these men who had designed the dual-purpose saddles for the mules, based on a saddle worn by camels in their native land. On the journey back from Balaklava to the siege lines around Sebastopol, the mules would carry shot and canister for the siege guns.

Alice had a mule to herself, but her clothes were heavily plastered with Crimean mud, the result of alighting frequently to tend to helpless men as the mules floundered through the pass which linked Balaklava to the inland regions of the Crimean peninsula.

The cavalcade had halted at Nell's Nest, where Alice had been able to produce tea and food for all those able to take sustenance, before accompanying them for the remainder of their journey to Balaklava.

However, for many soldiers the journey proved too much for constitutions weakened by months of privation and exposure to the Crimean winter, and, despite Alice's ministrations, by the time the cavalcade reached Balaklava many of the mule-back seats were empty, their late occupants buried in temporary, shallow mud graves, each marked by a rough wooden marker cross.

Alice looked tired, and Gideon realised there must have been many days like this. He wanted to hurry to her but, instead, held back as Alice helped the French colonial troops to lift men to the ground. Not until the last mule had been relieved of its human burden did he walk up behind Alice, unnoticed, and say quietly, 'Hello, Alice.'

She spun round and looked at him in total disbelief. For some moments her face registered a number of expressions, not least among them dismay, aware as she was of her dirty and dishevelled state. Then, as though they were meeting casually in an English village street, she said lamely, 'Hello, Gideon. You . . . I . . . this is a great surprise.'

Gideon had thought many times of what they would say to each other when they met, but nothing he had imagined had been quite as anticlimactic as this.

'Is that all we have to say to each other after all this time, Alice?'

Suddenly and quite unexpectedly, Alice was looking at Gideon through a veil of tears and she shook her head increasingly violently. 'No . . . No, Gideon, that isn't all. I've missed you. I've missed you so much . . . !'

Then Gideon was holding her to him and those soldiers who were well enough to be aware of what was happening set up a cheer. Watching from among the earlier arrivals, Nell nodded her head in silent satisfaction.

That evening, Ranald threw a party at his house to celebrate the arrival of Gideon and his reunion with Alice.

The principal guests were Gideon, Alice, Nell, Sailor and Bryn Roughley. There were also a number of Army Engineer officers, who had a professional interest in the proposed railway.

For Gideon and Alice it was a frustrating evening. They both wanted to spend time with each other, without others around, to reassure themselves that the love that had been kindled far off in Cornwall was still there.

Instead, as the evening wore on, more and more people came to the house, drawn by the promise of a party at which Nell and Alice were guests – Nell being known to many of the older officers, and the younger ones attracted to Alice.

The party was brought to an abrupt halt when, at its noisy height, a soldier entered the house, his cavalry uniform wet and muddy. He was quickly intercepted by one of the Engineer officers, and the two conversed for a minute or two before the Engineer turned to look round the room and pointed out Nell.

Observing the gesture, Nell met the soldier halfway across the room and he began talking animatedly. Nell listened, her expression serious, and she occasionally nodded. Then her glance fell upon Alice and she beckoned to the younger woman.

Gideon accompanied Alice to where Nell stood in conversation with the cavalryman. In response to Alice's query, Nell said, 'The Russian attack was more serious than we realised. Their cavalry swept through our lines to the rear of the trenches around Sebastopol, making for the gun emplacements. On the way they overwhelmed a compound containing women and children. The survivors have just arrived here, in Balaklava, and Doctor Forrest has asked for our help.'

Doctor Forrest was one of the most senior medical officers in the Crimea and he and Nell had worked together before, in China.

Turning to Gideon, Alice was clearly upset. 'I'm sorry, but I must go, Gideon. This is the reason Nell and I are in the Crimea.'

'You have no need to apologise, Alice. I'll come with you. There might be something I can do to help.'

Giving him a warm look, Alice reached out and took his hand, squeezing it gratefully, then, making hurried apologies for abandoning the party being held in their honour, she and Gideon put on their coats and hurried out of the house after Nell.

II

It was a pathetic group of women and children who huddled together on the quayside. Nearby, English soldiers, assisted by a number of French chasseurs, fought against rain, borne on a near-gale force wind, in a valiant attempt to stretch tarpaulins over an ill-matched framework of timber, in order to provide some shelter for them.

The chasseurs and a troop of British dragoons had escorted the unfortunate women and children from the siege lines to Balaklava. While Gideon helped the soldiers with the tarpaulins, Alice and Nell assessed the needs of the distressed families, some of whom were injured.

When the tarpaulin was satisfactorily secured, Gideon looked for Alice and saw her with a group of women. As he walked towards her, she suddenly broke off a conversation with a woman whose arm was heavily bandaged, and ran to where another woman was trying unsuccessfully to pacify a tearful young girl.

Dropping to her knees in front of the child, Alice reached out to grasp her by her upper arms. 'Verity! What are you doing here, poppet? Where's Mummy?'

Verity turned tearful eyes upon Alice but did not reply and the young woman with her said, 'Verity? Is that her name? You know her?'

'Yes – and Gwen, her mother, too . . . Where is she?'

'Probably dead by now,' the young woman said, adding callously, 'If she isn't she'll be wishing she was – and I've no doubt her wish will soon be granted.'

'I don't understand!'

'The horsemen who took us and our men by surprise were Cossacks. They slaughtered most of the men in the camp – some of the women and children too, but a few of the women were carried off. Someone said this kid's mother was one of those taken. I don't give much for her chances. But now I've found someone who knows the kid, I can leave her with you and go off and look for my husband. He was brought down here with the wounded, a few hours ago.'

'But . . . I can't look after her. Where's her father? He's a gunner, on the lines around Sebastopol.'

'Then, for Verity's sake, I hope he wasn't in one of the gun posts overrun by the Cossacks,' said the woman. 'They didn't leave any gunners alive. If he's dead then Verity's an orphan. If you don't want her you can do what I was going to do with her – put her on a boat to Scutari. That's where they gather the orphans before sending 'em back to England, ain't it? I've done my bit for her, now I'm off to find my husband. He'll be even more worried about me than I am about him, I dare say.'

Left with a still tearful Verity, Alice wondered what she could do to help her. It would be difficult to take care of her at Nell's Nest. Then she remembered the small boys and girls she had known in the Plymouth orphanages. She also remembered her promise to ensure Verity was placed in the care of her Cornish grandparents should anything happen to Gwen and her husband.

Even as she was pondering on the difficulties posed by the little girl, Verity bent down and a medallion tied about her neck on a piece of cord popped out at the top of her dress. On it was written Verity's name and, turning the wooden tag over, Alice saw carefully written, 'Next of kin Alice Rowe c/o Doctoress Nell.'

Tucking the medallion back inside Verity's dress, Alice thought it was an omen. A reminder of the promise she had made to Gwen.

'Hello, are you acting as a nursemaid for a while? I saw the little girl's mother hurrying off.' Gideon spoke to Alice, at the same time smiling at Verity.

'I wish it was as simple as that,' Alice replied. 'That woman was a stranger. Verity is the daughter of Gwen, a particular friend of mine. It seems Gwen might have been carried off by Cossacks who raided a British army compound up at the front.' Distressed, she repeated what the woman who had been taking care of Verity had told her.

'But Verity's father might be alive,' he pointed out.

'He might be,' Alice agreed, 'but the woman who had Verity didn't seem very hopeful. Until we know something, I'll need to take care of Verity. I'd better go and find Nell and tell her what's going on.'

Nell agreed that she and Alice should remain in Balaklava with Verity that night and take the small girl to Nell's Nest the next day. She would stay with them while they tried to ascertain the fate of Gwen and her husband.

They spent what remained of the night in Alice's old room in Ranald's house, where their arrival was the signal for the few remaining party-goers to finish their drinks and return to quarters in other parts of the town.

Gideon would be lodging in the house with Ranald for the foreseeable future, but after seeing Alice and Verity settled into their room he returned to the grounded *Argo* with Sailor and Bryn Roughley.

Work on the Crimean railway would not begin in earnest until the full force of navvies arrived in Balaklava with the remainder of their equipment. However, Gideon and the advance party would be busy doing what they could to prepare the route the line would follow to Sebastopol.

The first full day was a particularly busy one and Gideon did not return to Ranald's house until late that evening, to learn that Alice and Nell had departed for Nell's Nest earlier in the afternoon, taking Verity with them.

'Does that mean there was no news of Verity's mother or father?' Gideon asked Ranald.

'The only news we have is bad,' was the reply. 'A troop of cavalry came to Balaklava this morning to take the names of the women who survived the attack on the army compound. They were part of the force that pursued the Cossacks as far as they could. They succeeded in recovering the bodies of two British women, but neither has yet been identified. They also confirmed there were no survivors among the gunners in the posts that were attacked. Alice was very upset – and so am I.

246

Gwen stayed in this house for a while, with Verity, and Robert and I took her up to the front to find her husband. Gwen was – she *is* – a very nice young woman. I won't believe anything has happened to her until it's confirmed.'

Confirmation was received at Nell's Nest that Verity's father was among those killed by the Cossacks, but there was still no news of Gwen.

For the first few nights the little girl would wake screaming with fear and crying for her mother, but Alice had her sleeping in her room and, when she was so distressed, would take her into her own bed and cuddle her back to sleep.

When Gideon travelled to Nell's Nest two days later, Verity was being given a ride on a cavalry horse led by a young dragoon officer.

Standing nearby with Alice, Gideon said, 'It's very, very sad for Verity. What will happen to her now? You and Nell will have times when you have your hands full treating injured men – not to mention the trade you'll attract in this place.'

'I promised Gwen I wouldn't allow Verity to be taken to an orphanage,' Alice said determinedly. 'When I have a little time, later this week, I'll write a letter to Verity's grandparents in Cornwall telling them what's happened. I'm not absolutely certain of their address, but I'll just hope the letter reaches them. If they say they want her to come and live with them I'll arrange something.'

'Gwen chose wisely when she took you as a friend, Alice.'

'I don't think you choose friends,' Alice replied earnestly. 'They just happen.'

'It sounds like the way things are when you fall in love with someone,' Gideon said. 'You don't choose the one you're going to love. It just happens.' When Alice said nothing, Gideon asked, 'Isn't that the way it was for us, Alice? It certainly was for me.'

'We hardly knew each other before you left Cornwall, Gideon.'

'There's much we don't know, I admit, but I *do* know how I feel about you, Alice. You must feel very much the same about me, or you wouldn't have been so concerned that we might lose touch with each other.'

'I would have been very upset if we had never met again,' Alice agreed. 'But I want us to be quite certain – I want *you* to be certain . . .' Waving him to silence when he tried to speak, she went on, 'No, it needs to be said. I want you to be *really* certain, before we start making any serious plans for the future.'

Gideon's dismay was plain to see. 'But I thought . . . I . . . is there someone else, Alice?'

Alice gave him a reassuring smile. 'There was no opportunity to meet a man when I was with the Sisters of Mercy – or when I was working at Scutari. Besides, I've never *wanted* to find anyone else. I knew how I felt about you when we were both in Cornwall and that hasn't changed. I believe it's the way you want me to feel, but I need to know in my heart that *you're* certain. We've neither of us had the happiest of lives so far. I'd want that to change if we were married. I'd want us to be happy together for the rest of our lives.'

Gideon was both deeply touched and relieved by what Alice had said. She was not yet ready to fully commit herself, but she could hardly have made it more clear that she loved him too.

'I can wait, Alice – but I hope it won't be for too long. In the meantime I'll do my best to prove to you just how serious I am in the way I feel about you.'

18

I

The two gangs of navvies, together with the blacksmiths, carpenters, bricklayers and other artisans travelling on the *Argo*, were only the advance party of the total workforce required to build the Crimea railway, but they commenced working upon arrival – and worked hard.

Half the men stayed in Balaklava to unload the ship, while the remainder made their way inland to begin construction of a base camp. This would be important for them when they were too far away from the sea port to continue to use the facilities provided by the grounded *Argo*.

The camp would be virtually self-contained, and far better equipped than an army encampment. As well as an accommodation area it would include canteens, recreation facilities, a store, and various other comforts seldom provided for navvies. Petrie had seen at first hand the appalling condition of British soldiers in the Crimea. He was determined his navvies would not be obliged to suffer in a similar fashion.

Gideon had much to do along the route the railway would take, but he was able to spend half the nights of that first week at the base camp, close to Nell's Nest.

The doctoress's unique establishment was already being frequented by officers from Lord Raglan's headquarters, only a short ride away. Officers from the French headquarters also came to eat, even though they had a rather longer ride to reach it.

Alice helped in the kitchens when there were no casualties to be treated and Verity was never very far from her, although the small girl was now cared for by Hafise, the wife of one of Nell's Turkish waiters. A very beautiful young woman, Hafise had a nature that matched her face. She believed that a loving embrace was the answer to all the unhappiness of the world.

Entering the living quarters of the eating-house in search of Alice, Gideon watched Hafise for a while as she taught Verity to sing a Turkish song.

'She's an absolute gem,' Alice said, when Gideon made a complimentary remark about the Turkish nursemaid. 'Exactly what Verity needs right now.'

'You're very lucky to have found her,' Gideon agreed. 'But this war isn't going to last for ever. What becomes of Verity when the war is over and we all go back home?'

'I hope that long before that I'll have heard from Verity's grandparents and that they'll say they want her to live with them, in Cornwall.'

'What if they don't?' Gideon queried.

Alice gave Gideon a direct look. 'If they don't . . . ? Then I'll keep her myself. I feel I owe Gwen that, at least. But I don't think it will be necessary. I didn't know the Dymonds very well, but Parson Markham did, and he liked them a lot. I'm happy to trust in his judgement.'

'I'm quite certain they'll learn to love Verity very quickly, if they take her,' Gideon agreed. 'She's a very loveable little girl. There's not one of my navvies who wouldn't adopt her, given half a chance.'

'She's captured the hearts of most of the officers from Lord Raglan's headquarters too,' Alice said. 'Most bring a present of some sort for her when they come to Nell's Nest. If I'm not careful she'll be thoroughly spoiled. But I'm glad you feel that way about her too. Very glad.'

In spite of the magnitude of the task confronting him, the first few days in the Crimea were happy ones for Gideon. Nell ensured that no matter where her own presence was required, Alice remained at Nell's Nest. When Alice protested that the older woman was doing more than her fair share of the work, Nell pointed out that Verity needed her to be close by. In truth, she wanted Alice and Gideon to get to know each other once again.

However, little more than a week after the arrival of the *Argo*, more ships carrying stores and navvies began arriving from England, and Gideon had to return to Balaklava.

Alice was far more upset by Gideon's departure than she had anticipated. She realised that her feelings for him had become far stronger than they had been in Cornwall.

Watching her as she moped about the kitchen on the day after Gideon left, Nell said, 'You look about as happy as a wet holiday my girl. What is it, missing that man of yours?'

'Yes . . . yes, I am,' Alice admitted. 'I'm missing him far more than I thought I would. More than is good for me, that's certain.'

'There's nothing wrong with that if it makes you realise what he means to you,' Nell retorted. 'Far too many young folk waste precious time together, either out of misplaced pride, or because of what others might think is proper, or not proper, for them. You just follow your heart – and don't be frightened of letting him know how you feel. Life's far too short and unpredictable for young couples to play games with each other. I've lived a full life and all I look back on with regret are the things I never did when I found a man who really mattered to me. Believe me, Alice, if someone like Gideon had fallen for me, I'd have given up everything to follow him wherever he wanted to go.' Grimacing, Nell added ruefully, 'Come to think of it, I did it anyway – and for far lesser men.'

No fewer than twenty-three vessels were involved in the transport of men and materials to Balaklava for the Crimea railway.

Not all senior army officers were happy for civilian labour to be employed to make good the shortcomings of the military and it was not long before they began complaining that their needs in the tiny port were having to take second place to the requirements of the railwaymen.

Their complaints were made known in letters to London, but by the time they arrived the mishandling of the war had resulted in the downfall of the government of Lord Aberdeen. He had been replaced by the more energetic Lord Palmerston. Palmerston gave the railwaymen his wholehearted support and the subsequent telegraph messages from Robert Petrie to the Crimea made it clear that he expected Gideon and his men to justify the Prime Minister's faith in them.

Gideon needed no external pressure to build the railway in record time. He was becoming increasingly frustrated by the army's lack of organisation and distressed to see the number of men who paid for such inefficiency with their lives.

The usually happy-go-lucky navvies were also aware of the army's shortcomings. They realised that even an hour's delay in their work could cost the lives of many of their fellow country-men. As a result they were eager to break all records in the construction of the Crimea railway.

A friendly but nevertheless keen rivalry existed between the gangs working on the railway. A double line was being laid between Balaklava and the siege lines and Gideon put the gangs of Sailor and Bryn Roughley together on one line, setting the pace for those working on the other.

The navvies rose to the challenge, and by the end of the first full day of track-laying the railway had advanced inland well beyond the narrow confines of Balaklava and its harbour. It was a pace they maintained during the days ahead and, before long, the efforts of the navvies were making a significant difference to the soldiers besieging Sebastopol.

In a matter of days they had conquered the quagmire of the pass which in the earlier winter months had almost succeeded in isolating Balaklava from the interior of the Crimea. Now supplies and ammunition could be carried through the pass on railway wagons which returned with those requiring urgent transport from the front.

It was only a beginning, but to thoroughly exhausted men it was sufficient to make the difference between life and death.

Senior army officers came to watch the navvies at work and reluctantly marvel at their energy. Many had been strongly opposed to their presence in the Crimea – especially as they would not be subject to military discipline.

They saw nothing to change their opinion about the navvies' lack of discipline, but even the sternest critic went away filled with a grudging admiration for the manner in which they worked. One dour infantry colonel declared that if the navvies could be persuaded to don uniform and storm Sebastopol, the city would fall in the course of twenty-four hours!

II

The railway reached Nell's Nest, where the depot had been established, almost a month ahead of schedule and it was a feat that Nell celebrated in her usual flamboyant manner. She invited them all to a feast they would talk about for the remainder of their lives.

It was also a shrewd move on her part. Fully replete, each navvy left the eating-house vowing to return for one of her meals whenever his work permitted – and the navvies had far more money to spend than most of the soldiers who found their way there.

In fact, as Nell was aware, although the line would not be completed for another month or six weeks there was already good cause for celebration. It was little exaggeration to say that every sleeper put in place on the railway line represented a soldier's life, saved as a result of slashing the time taken in transporting him to medical facilities at Balaklava, or on to Scutari.

Because the work was being carried out in occupied enemy

territory, with much of Russia's vast army only a short distance away, it was inevitable that details of the railway's progress would be known to the Russian army commander. Equally inevitable that he should try to bring the work to a halt.

The morning after Nell's party, the navvies were assembling at their depot a little later than usual, ready to be transported to the railhead, when they were startled to hear the sound of small arms fire from not very far away. It was accompanied by the noise of much shouting.

'What d'you think it is?' The question came from Sailor who was standing talking to Gideon.

'I don't know – but we'll take no chances. Here's the key to the ammunition wagon. Open it up quickly and get the revolvers out. I'll round up some of the men and send them to help you.'

The closed wagon containing the revolvers was always kept wherever the bulk of the navvies happened to be, and, from the day they had commenced work, Gideon had detailed two men experienced in the use of firearms to check the weapons each morning.

Every Sunday, when the navvies had a rest day, the same two men would empty the chambers of each revolver and reload them, using fresh powder. Ranald had grumbled that it was an unnecessary expense, but Gideon had given care of the guns and ammunition to Sailor who had once been a naval gunner. It was he who insisted that such precautions were necessary, pointing out that it was useless having the weapons if they weren't ready to fire at a moment's notice.

His precautions were about to be justified.

Not ten minutes after the sound of shooting was heard, a young trooper came into view. Galloping his horse towards them, he pulled it to an untidy halt before Gideon. With breathless urgency, he said, 'Excuse me, sir, are you in charge here?'

'Yes. What's happening? What was the shooting we just heard?'

'It's a squadron or two of irregular Cossack cavalry, sir. They must have slipped through our lines during the night hours. Lieutenant Penhaligon and our troop were on dawn patrol when we met them. They're heading this way and Lieutenant Penhaligon says he has no doubt you're their target. He says you're to make for headquarters. We have enough infantry there to beat them off and he's sent another trooper ahead to alert them. In the meantime he'll try to hold the Cossacks back for as long as possible.'

Even as he spoke there was another outbreak of sporadic gunfire, this time closer than before.

'How many Cossacks are there?' Gideon demanded.

'There must be almost three hundred, sir . . . but you must get moving, quickly!'

'How many men in your troop?' Gideon persisted.

'Twenty-four, sir . . . no, twenty-two. I'm here, and Trooper Judd was sent to warn headquarters. But you must get moving . . .'

'We're not going anywhere,' Gideon said firmly. 'We can't afford to lose our stores, they're far too important. We have revolvers and I'll distribute them among the men. The main store buildings are iron-clad so we'll fight the Cossacks off from them. Gallop back to your lieutenant and tell him to get his troop here as quickly as he can. There's no sense in sacrificing good men. With their support we can hold the Cossacks off until help arrives.'

The trooper hesitated uncertainly, but Gideon had already turned away, calling, 'Sailor! Get the guns and the men into the storehouses . . . Hurry now! Russian Cossacks are on their way here.'

Sailor immediately stopped handing out revolvers and detailed navvies to take the boxes containing them into the storehouses, together with all the ammunition they could carry.

The trooper wheeled his horse and galloped back the way he had come. As he did so, Gideon saw Alice standing in the doorway of Nell's Nest.

Calling to his other senior ganger, Gideon shouted, 'Bryn, take a couple of navvies with you and get everyone from Nell's Nest into the nearest store . . . right now! The rest of you help me push these wagons clear of the stores to give us a clear field of fire.'

For some minutes the railway depot was the scene of fren-
zied activity. Trucks were pushed away from the storehouses
and Alice, carrying Verity in her arms, together with Nell and
the mainly Turkish staff were hurried to safety inside the stores.

They had only just been bundled inside when the young
trooper who had delivered the warning unexpectedly galloped
back to the depot. He was alone.

Sliding his horse to a halt, he called, 'Get inside right away,
sir. The Cossacks are close behind me.'

'You heard him,' Gideon shouted. 'Everyone inside the stores
– and barricade the doors. Shoot from the windows as soon as
the Cossacks are close enough but, remember, you only have
revolvers and they have a short range.' Turning back to the
trooper, he said, 'You'd better get inside too and take your horse
with you – but where are the rest of your troop?'

Close to tears, the young trooper replied, 'There was no sign
of them, sir. All I could see were Cossacks, lots of them – here's
the first of them now . . .'

The depot area had been levelled by the navvies. The land
beyond was broken country with gently sloping hillocks and
shallow depressions. It was from here that horsemen mounted
on wiry ponies were emerging.

'Get inside . . . Now!' As Gideon shouted the order and ran
for the doorway of the nearest store he heard a shrill scream of
fright. It came from Verity, in the store closest to Nell's Nest.
She had seen the Cossacks and it brought back terrifying memories for her.

There were three iron-clad stores and Gideon ran for the
centre one of the three. The windows were fitted with stout iron
bars to keep out thieves and also had sturdy iron shutters, but
these were swung back and fastened to the outside walls during
daylight hours.

Satisfied the doors of the other two stores were secure, Gideon
helped to swing the heavy door of the centre store closed. They
were only just in time. As the iron door clanged shut there was
a ragged volley of musket fire from the oncoming horsemen
and musket balls struck the door and walls of the store, some

ricocheting off again, leaving an incongruous musical sound in their wake.

A couple of the navvies shot back at the Cossacks, but Gideon called for them to hold their fire. The horsemen were well out of revolver range. Few of the navvies had ever fired a gun before and much ammunition would be wasted, even when the Cossacks were close enough to be hit.

The trooper who had raised the alarm stood beside Gideon, watching the Cossack horsemen from a window in the centre store. Raising his carbine and aiming carefully, he fired as they slowed uncertainly before the stores.

A Cossack slumped forward on his horse's neck and the trooper grunted in satisfaction, saying, 'We're lucky they're irregulars, sir. Not many have muskets. The majority are carrying only a sword, or a lance.'

Few of the Cossacks seemed aware that one of their number had been shot. Even had they known, it was doubtful whether their tactics would have changed.

Shouting wildly, they charged towards the buildings, as though expecting them to crash to the ground before them. The store in which Alice was sheltering was the first to receive their attention but the horsemen were greeted with a fusillade of shots from within and a number of Cossacks fell to the ground, as did a couple of horses.

A second fusillade achieved a similar result. Although some of the Cossacks returned the fire, many more moved back out of range.

It was now the turn of the navvies in the other two stores to open fire upon the horsemen and the Cossacks suffered more casualties.

But the navvies were taking casualties too now. Two men in Gideon's store received wounds, despite being on the far side of the building from the window through which the shots were fired. One navvy was hit in the upper arm, the other had a musket ball lodged in his shoulder.

Thwarted in their aim of destroying the storehouses, the Cossacks retired to a safe distance to decide on their next move.

They had been unaware the buildings were iron-clad and, in any case, had expected to take the navvies by surprise.

After a hurried discussion, the Cossacks turned their attention to Nell's Nest and the nearby living quarters of the navvies. Nell was sheltering in the same storehouse as Alice, but when she realised what was happening her howl of anger could be heard in all three stores. The Russian horsemen began looting everything that could be carried away with them and wrecking much that remained.

Fortunately, thanks to Robert Petrie's generosity, much of Nell's Nest had been constructed of similar material to his storehouses. As a consequence little permanent damage was caused to the building itself. The navvies' living quarters were also iron-clad and the doors heavily padlocked. Although the Cossacks succeeded in forcing one building open, the locks on the others held.

However, there was loot enough in Nell's Nest to delight the Cossacks. They carried food and cooking utensils from the building and even tried to mount their horses carrying chairs in their arms. The trooper with Gideon succeeded in downing two of the looters with his accurate shooting, but the property they had been carrying was promptly taken up by others.

The navvies in the nearest store to the eating-house kept up a fusillade of revolver shots in response to Nell's outraged cries, but they were too far away from Nell's Nest to cause any serious wounds among the looters.

Suddenly, the watchers became aware that those Cossacks who had dismounted were running for their horses and a sense of urgency had suddenly come over the Russians.

The navvies in Gideon's store could see nothing, but when cheering broke out from the store with windows facing in the direction of the British army headquarters, the young trooper cried, 'It's our men! Trooper Judd must have got through!'

In fact, it was not British troops but French cavalry coming to the rescue of the navvies and those from Nell's Nest, but the result was the same.

Nevertheless, the Cossacks were defiant even in retreat. They fired upon the storehouses as they swept past them – and one of the horsemen threw a lance, much as if it were a javelin, which came through the window of one of the stores, scattering the defenders.

But it was a musket ball, ricocheting from one of the iron bars at a window, that scored the most notable victory for the retreating Cossacks.

Standing to one side, his back to the window, Gideon was laboriously reloading his revolver when the deflected musket ball struck him, knocking him forward.

He fell to his knees, the revolver dropping from his hands, a fierce pain in the upper part of his back. Then he fell face forward to the ground and lay there fighting for breath.

III

Nell and Alice were binding the wrist of a wounded navvy when the call went up for Nell to go to the centre store.

'I'll be there as soon as I've finished what I'm doing,' she replied. 'What's happened?'

'It's Gideon,' said a navvy who had come from the centre store. 'He's been shot in the back by the Cossacks.'

Alice was running from the store before the man finished talking. The navvy Nell was bandaging pulled his arm away from her, saying, 'I'll get someone else to tie this. Go and help Gideon.'

Alice found Gideon lying on his stomach, his face turned to one side, breathing laboriously. Holding back her tears, she dropped to her knees beside him. By the time Nell arrived and pushed her way through the crowd surrounding them, Alice had placed her own coat beneath Gideon's head and he was trying unsuccessfully to reassure her that he would be all right.

'Stand back and give me some room,' Nell said to the worried navvies. 'Alice, you know where my wound dressings are kept. Go and see whether the Cossacks have left us any. There's a jar marked "antiseptic" with them. Bring that to me as well. In the meantime . . .'

She looked about her and spotted one of the dozen or so French cavalrymen detached from the main body pursuing the Cossacks and told to remain at Nell's Nest to give whatever help might be required.

Speaking to him in passable French, she said, 'Do you know Surgeon Larrey, at your headquarters hospital?'

'I know of him,' the Frenchman replied.

'Go to him as quickly as you can. Tell him Mother Nell has urgent need of some chloroform. Bring it back to me right away.'

Impressed that Nell should know the army surgeon, the cavalryman hurried away to do her bidding. Surgeon Larrey was not only the most senior of the French doctors, he was also very highly regarded by the men he treated.

'What are you going to do?' The question came from Sailor, who was extremely worried about Gideon.

'There's no exit wound,' Nell explained. 'It means the musket ball is still inside him. Judging by the way he's breathing it's close to a lung and needs to come out as quickly as we can get at it. Fortunately, it doesn't appear to have damaged the lung, or he'd be breathing blood, but I have to locate the ball and it isn't going to be easy. In fact, unless he's unconscious it will be impossible – and very dangerous. That's why I need chloroform.'

Turning to the navvies, who had been listening closely to all she had been saying, she asked, 'Is there anyone here who can tell me what Gideon was doing when he was shot?'

A number of the navvies claimed to know, but each told a slightly different story. It was the cavalryman who had first given the warning of the marauding Cossacks who eventually said, 'I know exactly what he was doing, Mother Nell. I was holding the percussion caps for him while he loaded his revolver.'

'Good! I want you to show me exactly how and where he was standing when the musket ball hit him.'

Puzzled, the cavalryman self-consciously moved to where Gideon had been standing. Bending over he adopted Gideon's posture at the time he was shot. Then he straightened up, but Nell ordered him to remain stooped over, loading an imaginary revolver.

When she was satisfied, she went to the window and, turning back to the navvies, asked, 'Did anyone suffer a near miss when the shot came through the window?'

The men looked at each other, puzzled by the question. Then one of the Welsh navvies said, 'Yes, I did. It damn near took off the tip of my nose.'

'Come here and show me where you were standing,' Nell said unsympathetically.

'What for?'

'You don't need to know why, just do it,' said Bryn Roughley, beginning to understand why Nell was asking such questions.

The navvy moved to the window and looked out before shifting his position slightly. 'I was standing here, just like this, see?'

'And the bullet narrowly missed your nose, you say?' Nell began examining the iron bars at the height of his nose and suddenly exclaimed, 'Ah! I think we have it. This looks as though a musket ball has glanced off the bar, just here. Right. Someone get a long piece of string.'

A length of string was produced and Nell ordered the man who had suffered the near miss to hold one end of the string to the mark on the window bar. Keeping the string taut, she then took it to where the puzzled cavalryman still held his pose and placed the string on his back in a spot identical to Gideon's wound.

Then, with Bryn Roughley holding the string in place, she stood back and looked critically from the string to the back of the cavalryman, estimating the angle at which the musket ball must have entered Gideon's back.

When Alice returned to the store Nell was crouching beside Gideon, trying to reassure him that all would be well. When she stood up to take the antiseptic and dressing from Alice, the younger woman asked anxiously, 'What are you going to do with him, Nell?'

'I'm going to hurt him,' was the apparently callous reply, although Nell said it too quietly for Gideon to hear. 'But first I need to clean my hands with this.'

Taking the top off the jar, she poured some of the antiseptic over her hands and cleaned them thoroughly. When she kneeled beside Gideon once more, Sailor asked, 'Is there anything we can do to help, Nell?'

Before she could reply, Gideon, speaking through clenched teeth, said, 'You can do something to help . . . me, Sailor. You and Bryn get work started again. We can't afford to waste time . . . for anything.'

Sailor looked at Nell uncertainly and Gideon said, more forcibly, 'Get on with it, Sailor. You and Bryn are in charge now. Don't let me down.'

The effort of speaking so vehemently left him coughing. When the spasm eased off slightly, Nell said, 'They can go in just a few minutes, but first I have need of them.' As she spoke she ripped Gideon's shirt from his back, peeling it away from the wound. Addressing Alice, she said, 'If you don't want to help, turn your head away, girl. You won't like this any more than he will.'

Removing the temporary dressing that had been placed on Gideon's wound to stem the bleeding, she unceremoniously poured more of the antiseptic liquid on to the wound and the surrounding area. Gideon shouted in agony, but as his body shuddered with pain she said to Sailor and Bryn Roughley, 'The two of you hold him down. Keep him as still as you can. What I'm about to do to him will hurt like the devil's fork, but it needs to be done.'

As the two men did as they were told, Nell said to Alice, 'I'm sorry, but I'm going to need you, Alice. I want you to hold his head. Hold him as though he's the most precious thing in the world.'

Crouching over Gideon's head, holding it to her, Alice knew she should have turned her face away, as Nell had suggested. Instead, she watched what the other woman was doing and was horrified to see her insert her finger into the wound, ignoring Gideon's cries of pain.

Nell's exploration of the wound was crude, but it was also quick and efficient, even though Gideon was left breathing as though he had just been running a gruelling race.

With tears running down her face, Alice cried, 'Why did you have to do that, Nell? I thought he was going to die.'

'As well he might, had I *not* done it,' Nell explained. 'He's got a musket ball in there, lying against his lung – and a fragment of rib too, if I'm not mistaken. I needed to find out exactly where they are, and how close to his lung. They've got to be taken out, Alice. If we leave them and the lung becomes infected – or perhaps perforated . . .'

She did not complete the warning, but Alice was left in no doubt about the outcome of such an eventuality.

'But . . . is it something you can do, Nell? Have you ever taken a musket ball out of a man before?'

'Once or twice,' Nell said, 'but never from quite such a dangerous place as this.'

Wide-eyed, Alice asked tremulously, 'Do you really think you should attempt it, Nell? What if . . . what if something goes wrong? No, you can't take such a risk. Not with Gideon.'

'Do you think I *want* to do it?' Nell spoke far more sharply than she had intended. Aware of it, she said, more gently, 'You've seen some of our army doctors at work, Alice. Would you rather trust Gideon to them than to me?' When Alice made no reply, Nell went on, 'I told you I haven't removed a musket ball from so close to a man's lung before, Alice, and that's the truth – but I doubt whether more than one or two army doctors have done it, either. Even so, for a great many years I've been doing everything a surgeon does, and doing it as well, or better, than any of 'em. I know what's inside a man's body, even though I'll never tell anyone what I needed to do in order to find out. All I will tell you for certain is that if that musket ball and the piece of bone that's in there with it don't come out, then Gideon's not going to pull through. One single bout of coughing could be enough to finish him. I'll get Sailor to fix up an operating table over by the window while I go off and look for my forceps and surgical things. In the meantime, think about what I've said. When I come back, you can tell me what you want me to do – or not do.'

IV

Nell's skills – or the lack of them – as a novice thoracic surgeon were never put to the test. When she went in search of her case of surgical instruments, she discovered it had been stolen by the marauding Cossacks.

Another appeal went to the senior French army surgeon, this time in the form of a note, explaining what had happened and telling him what she intended doing. She asked the surgeon if he would be kind enough to loan, or donate, a set of surgical instruments in order that she might carry out the operation.

Surgeon Jacques Larrey supplied Nell with the chloroform and surgical instruments she had requested, but instead of sending them with the French cavalryman he brought them to Nell's Nest himself.

After greeting Nell warmly, he asked to see Gideon, who had been taken to Alice's room pending the arrival of the surgical instruments.

On the way, Nell explained how Gideon had received his wound and described what she had done to locate the musket ball. The surgeon nodded approvingly. 'You have lost none of your skills,' he said. 'But you have admitted you have never before performed such an operation. I have. I will operate on this young man and you will assist me. Then you will know what to do in any similar emergency in the future.'

Alice was sitting with Gideon, who was conscious once more, but after greeting Alice and chatting briefly with Gideon Surgeon Larrey suggested that Gideon should be moved back to the storeroom where Nell had decided the operation should take place. There was plenty of light from the windows and a long, narrow table had been vigorously scrubbed and put in position.

When Gideon was carried to the store, Alice wanted to stay and help during the operation, but Larrey, aware of her relationship to his patient, insisted, gently but forcibly, that she occupy herself elsewhere. He explained that, in such a delicate operation as he felt this would be, even the slightest sound from her could cause sufficient distraction to affect its outcome. He suggested she go away and prepare the room where Gideon would be taken to convalesce after his work was done.

Larrey's talk of convalescence was reassuring, but Alice had learned enough about nursing to realise that the operation upon Gideon would require considerable skill – and a great deal of luck.

It had been decided that Gideon should have Alice's room while he was recovering, during which time Alice would have a bed in Nell's room. Alice kneeled beside the bed she hoped would soon be occupied by Gideon, and prayed as never before that all would be well with him. Then, determined to keep herself occupied, she began cleaning and tidying.

As she worked, Alice thought of all that had happened since Gideon had come into her life. Many things had tested their love for each other, yet there seemed to be a pattern to it all, a convergence of their two lives. She tried to convince herself that a chance shot from a Cossack musket could not end it all in such an abrupt and tragic manner.

Alice worked hard, believing that the operation was likely to take a long time. However, far sooner than she expected, she looked up to see Nell standing in the doorway.

It was so quick that Alice immediately feared the worst. 'Nell! What's happened? Gideon's not . . . ? He's not . . . ?' She could not bring herself to say the word.

'He's not anything,' declared Nell. 'And I've just seen an exhibition of skilful surgery that I doubt I'll ever see repeated. Instead of probing for it, Surgeon Larrey estimated where the musket ball would be and cut between two of Gideon's ribs to get at it. He was so accurate, it was uncanny. Not only was he able to remove the ball in a matter of minutes, but he also removed a fragment of rib – and a piece of Gideon's shirt that was in there. It was all done so quickly and efficiently that Gideon hardly had time to bleed over him.'

Alice winced at Nell's graphic report. 'How is he, Nell? Is he going to be all right? Can I go and see him?'

'He's breathing far more easily and Surgeon Larrey says there's no reason why he shouldn't make a full recovery, although he'll be sore for a while, I dare say. As for seeing him ... He's still unconscious from the chloroform but he'll be carried in here in a few moments. I suggest you get the bed ready for him.'

Gideon was brought to the room on a stretcher carried by two navvies. They had not heard Nell's optimistic report to Alice on the success of the operation and, unfamiliar with the properties of chloroform, were concerned because Gideon was unconscious and not moving. Lifting him carefully from the stretcher they laid him face down on the bed, and Alice could see the dressings covering his wound and the site of Larrey's incision.

'Is he going to be all right, Mother Nell?' one of the navvies asked.

Surgeon Larrey was just entering the room and heard the question. 'My operation was a complete success,' he assured them. 'As for his recovery . . . that is in the hands of Mother Nell. But she is such a skilful healer that I would entrust my own life to her.'

'There you are, boys,' Nell said, 'you can go back and tell the rest of the men what Surgeon Larrey has said. God willing, Gideon will be back with you in a week or two.'

When the two navvies had left the room, satisfied that Gideon was going to recover, Alice said to Larrey, 'Is Gideon really going to be all right?'

'Would I put my reputation at risk by telling an untruth about one of my patients?' the surgeon replied. 'You can see for yourself his breathing is already much improved. However, perhaps it would not be good for him if yours was the first face he saw when the effect of the chloroform wears off.'

'Why?' Alice succeeded in conveying both dismay and bewilderment in the single word.

'Because one look at a face such as yours and he will surely believe my operation was a failure and that he has gone to heaven.'

Nell smiled. 'Pay no attention to him, Alice. Jacques Larrey is a true Frenchman, as skilful with flattery as he is with surgery.'

'When it is the truth it is gallantry, not flattery, Mother Nell. Did I not tell the two men who were just here that you are the finest healer I have ever known? Tell me, was that flattery, or was I speaking the truth?'

'You are a fine surgeon, Jacques, and one of my favourite men. Give me a few days to get things sorted out here, then bring some friends along with you to Nell's Nest for dinner one evening. I promise you a meal you won't have tasted the likes of since leaving France.'

'It is an invitation I accept with very great pleasure,' said Surgeon Larrey. 'But now I must return to my duties with the French army. Call me again if you feel your patient has need of me. If not, I will look in on him when I come for that special meal, but I believe that, thanks to your prompt actions, he will recover quickly.'

Inclining his head to both women, Surgeon Larrey turned to go, but Alice called to him and when he turned back she ran to him and kissed him warmly on the cheek. 'Thank you. Thank you very much.'

Touching his cheek, the French army surgeon smiled. 'An invitation to a Mother Nell meal, and now a kiss from a beautiful young woman. This is the first day I wish to remember since I arrived in the Crimea. Thank you both.'

V

Happily, Gideon's recovery was as rapid as Surgeon Larrey had predicted. Although still uncomfortable, he was out of bed in five days and discussing the work of the navvies with Ranald MacAllen each evening.

He was anxious to get to the advancing railhead, to check for himself how the work was progressing, but Nell would not hear of it. She was adamant he would not be making such a journey until she was satisfied it would not set back his full recovery. She threatened that if he attempted to disobey her, she would take away all his clothes to prevent him from leaving his bed.

During his convalescence, one of Gideon's most constant visitors was Verity. She would sometimes accompany Alice to his room but, more often, would be brought there at her own request by Hafise, her Turkish nursemaid.

Gideon had had very little to do with young children since going out into the world on his own, but he was amused by Verity's unselfconscious chatter and found himself becoming increasingly fond of her and looking forward to her visits. When he was well enough to stay out of bed for most of the day, and even wander round Nell's Nest, Verity would often keep him company, with the Turkish nursemaid never far away in case he should find Verity's constant chatter too tiring.

After a while, Sailor and Bryn Roughley would accompany Ranald to Gideon's room for the evening 'briefing'. Although Gideon would not be happy until he was once more actively involved in the work, it was apparent that the railway was advancing well ahead of schedule.

Once it became clear that Gideon was on course for a full recovery, Alice was very happy to have him staying at Nell's Nest. She had fussed over him when he was too ill to do anything for himself, and now he was a little better she chided him for trying to exceed his capabilities, but she was happy to accompany him when he was able to walk for short distances about the depot.

Alice no longer worked in the restaurant because Nell had acquired a number of helpers there. Among them were a couple of artillery widows whose husbands had been killed in the raid on the gun emplacements about Sebastopol, when Jack Dymond died and Gwen was carried off. The women could have been repatriated to England, but the army was in their blood and they would remain in the Crimea and probably marry again.

Nell also employed a number of soldier's wives from the nearby army headquarters, on a part-time basis.

Alice's main task now was to help Nell treat the soldiers and officers who came to Nell's Nest seeking her patent medicines and ointments for a variety of ailments. She also tended the weak and wounded soldiers who passed through on their way to Balaklava for passage to Scutari, or perhaps to England.

Despite such duties, Nell ensured that Alice was able to spend as much time as possible with Gideon and was gratified to observe that the couple had become very much closer to each other as a result.

Alice was aware of this too. She found herself thinking of Gideon for much of the time when she was not with him and looked forward to their meetings with an eagerness that increased with each passing day.

One day, Gideon was in his room watching through the window with Verity as Ali fed the fifty or so chickens that Nell had somehow acquired and securely penned close to the main building of her eating-house to prevent them from being stolen. Verity was clapping her hands with delight at the greedy antics of the chickens when Alice entered the room unexpectedly. She would normally have smiled to see Verity so happy, but there was no smile today and no humour in her expression.

When she spoke, her words were not for Verity or Gideon, but for the Turkish nursemaid, who was tidying the room while Gideon entertained the little girl.

'Hafise, why don't you take Verity outside and see if Ali will let her feed some of the chickens?'

Realising that Alice wanted the child out of the room, Hafise stopped what she was doing, gathered her up and led her outside to 'help Ali'.

Gideon too was aware there was a reason for Alice's request. 'What is it, Alice? Have you had some bad news?'

'Yes . . . It's about Gwen. Some wounded Cossacks came down the line this morning. They'd been taken prisoner after the raid on Nell's Nest but were only now fit to travel. Nell, me, and one of the Russian-speaking waiters took food and drink out to them when they stopped here. The waiter told Nell when he heard one of the Cossacks saying he wished they'd found a woman like me in the family compound when they attacked the gun positions around Sebastopol. With the help of the waiter, Nell and the officer in charge of the escort questioned the prisoner, who admitted that the Cossacks had carried off some of the women. Of course, he claimed that he had nothing to do with it personally. But he said they took six women away with them when the fighting was over. Two were killed before they got back to the Russian lines. Another two died of "ill-treatment" and were thrown into the River Tchernaya. The other two women were carried back to the Cossack camp where one died a few days later.' Alice's voice broke as she added, 'He said the last remaining woman was ill-treated too, but she was taken to another camp. He never saw her again, but later heard that she'd died as well.'

She was deeply upset and Gideon crossed the room and put his arms about her. When she was a little more composed, Alice tried very hard to control her voice as she said, 'Our soldiers found the bodies of the first women to be killed, but Gwen wasn't one of them. I hope . . . I hope she was one of those who were thrown into the river and didn't survive to be one of the last two.'

Holding her head close to his chest, Gideon said, 'Try not to think about it, Alice.'

As he spoke he could see Verity through the window. She was squealing with delight as chickens clucked about her, pecking at the tiny handfuls of corn she was being allowed to distribute.

'I think it's so sad that a lovely little girl like Verity should be left without a mother or father to take care of her,' he said. 'Still, she at least has her grandma and granddad.'

'She has me, too,' Alice said fiercely. 'I'll see to it that she doesn't grow up with no one special to love her.'

Remembering the hard life Alice had led, Gideon held her

closer still and said, 'That's something everyone should have, Alice, no matter what age they are.'

'Yes.' Alice was silent for some minutes before she looked up at him and said, 'I'm glad we found each other again, Gideon. Very, very glad.'

VI

A fortnight after being wounded, Gideon was back with his navvies. Meanwhile, the railway had edged ever closer to the siege lines around the Russian port of Sebastopol, allowing nothing to delay its progress. Cuttings were dug, bridges built – sometimes in a single day – and the railway advanced inexorably towards its goal.

Although the line was being constructed at a phenomenal rate, the builders still had their critics among the senior officers, who constantly complained about the lack of discipline among the hard-working, hard-drinking and occasionally hard-fighting navvies. For their part, the navvies were contemptuous of officers who seemed content to command ill-equipped soldiers who were required to live, fight and die in atrocious conditions, while the staff officers had comparatively comfortable quarters. Indeed, Lord Cardigan, commander of the ill-fated Light Brigade, enjoyed positive luxury, dining and sleeping on board his own luxurious private yacht, anchored off Balaklava.

The navvies found it difficult to comprehend the fact that many of the officers looked upon the war as a form of al fresco entertainment. When a battle was imminent, their wives, families, friends and other visitors from England would take picnic baskets to vantage points to watch as men fought desperately for their lives. The onlookers would occasionally take a convalescing officer along with them, to give a professional commentary on what was happening.

It was a far cry from the condition of the infantrymen who were suffering dreadfully as a result of disease and exposure to the Crimean winter. Although the railway was slowly easing their lot, almost half the British army in the Crimea was incapacitated. Those on the siege lines about Sebastopol – referred to as a careworn, threadbare and ragged army – were forced to rely upon the goodwill of the better organised French army for their very existence.

The railway would make a great difference to the situation by providing a dependable means of transport. Food, equipment and ammunition could be relied upon to reach the siege lines and the trains represented a rapid means of evacuation to Balaklava for men exhausted by the conditions under which they were forced to live. In addition, more of the men could be sent direct to Balaklava, where some of Florence Nightingale's nurses from Scutari were now stationed, instead of being forced to remain in hospitals close to the siege lines, where standards were far from satisfactory.

However, as Nell had said on many occasions, the place where most lives could and should be saved was at the front itself, where medical facilities, if they existed at all, were at their most primitive. Nell would often set off alone from Nell's Nest carrying a satchel of medical supplies slung over her shoulder. Saying only that she was going to visit the soldiers manning the siege lines around Sebastopol, she declined Alice's offers to accompany her. Until the railway track was nearing the siege lines those who worked at Nell's Nest did not realise what she actually did there.

One morning, when Gideon and the navvies were setting off to the end of the line, Nell, dressed in her usual garb of yellow frock, heavy shawl and red-flowered hat, asked if she might ride with them in order to pay one of her visits to the soldiers. When the workers' train reached its destination, Nell set off towards the trenches while Gideon and the navvies set about their work.

Mid-morning, when the job was progressing satisfactorily, Gideon left Sailor and Bryn Roughley in charge and made his way to the British siege lines that overlooked the heavily fortified Russian town. He was going to see the colonel in charge of artillery, who had his headquarters in an observation post among the trenches. It was necessary to discuss with him where the small branch lines of the railway should be laid, in order to best serve the front-line gun emplacements.

On the way, Gideon heard the occasional exchange of small arms fire, but took little notice. Both sides were in the habit of shooting at anything that moved in enemy-held territory. As often as not, shots would be fired randomly into the other's lines, aiming at nothing in particular.

A fresh outburst of firing occurred when Gideon reached the British lines, but, dropping into a trench, he made his way along it, heading in the general direction of the artillery colonel's head-quarters. He had almost reached his destination when he came to a section of the line where the men in the trenches were extremely agitated.

'What's going on?' he asked an Irish Fusilier sergeant who appeared to be in charge of this section of the trenches. 'Are the Russians up to something?'

'You could say that,' the sergeant replied. 'They usually are – but right now we're more concerned about Mother Nell.'

'Mother Nell?' Gideon was startled. 'What's she doing?'

'The Russians attacked one of our forward picquets, while Mother Nell was here. Our men managed to beat them off, but a couple were wounded in the process. Before we knew what she intended doing, Mother Nell climbed out of the trench and ran down to the picquet's position. I think the Russians must have been taken as much by surprise as we were because they didn't fire a shot until Nell reached the picquets – then all hell broke loose for a while! I hope she stays there until nightfall. I doubt if she'll be so lucky if she tries to come back.'

The sergeant's suggestion that she should stay with the men in the forward post made sense – but Nell had other ideas.

The firing from the Russian lines had died down and the sergeant was instructing Gideon on the best route to take to reach the artillery observation post when one of the Irish soldiers shouted in disbelief, 'It's Mother Nell. She's coming back, sergeant – and she has Tom Casey with her.'

The sergeant and Gideon raised their heads above the parapet of the trench and saw Nell in her colourful garb, struggling through the mud of no-man's-land, a hefty arm about a soldier who had a bandage about his head and seemed to be having great difficulty staying on his feet.

As the men in the trench watched, the Russians began firing from their own lines. Gideon could see where musket balls were hitting the ground around Nell and the soldier, sending up small angry splashes of mud and dirty water. Suddenly, as the men watched, Nell lost her footing and fell down in the mud, taking the wounded soldier with her.

Nell was a heavily built woman and she struggled in vain to rise to her feet. Her attempts were severely hindered by the heavy satchel she carried. It appeared to be trapped beneath the fallen soldier. Shots were still being fired at her, but the sergeant climbed over the parapet of the trench to go to Nell's aid. Gideon went with him.

When they reached Nell the sergeant quickly freed the satchel. Then, with some difficulty, he and Gideon pulled Nell to her feet, by which time two more soldiers had come to their aid and lifted their wounded comrade.

The Russians continued firing at them until Gideon, Nell and the soldiers dropped down into the British trench. By some miracle, not one of the party was hit, although Nell later discovered a spent musket ball inside her satchel.

'What on earth did you think you were doing out there, Nell?' a concerned Gideon demanded. 'You could have got yourself killed.'

Nell was muddy and cross and her treasured hat was askew, but drawing herself up to her full height she looked fiercely at Gideon and retorted, 'What are *you* doing here? You're hardly fit enough to walk, let alone come up here in such conditions.'

Gideon's jaw dropped in disbelief. 'You tell me *I* shouldn't be . . . ! Nell, you were being shot at out there. You came close to death. *You're* the one who shouldn't be here.'

Apparently unconcerned, Nell took out one of her small cheroots and struck a match to light it. Looking down at her muddy clothes, she said, 'Look at me. This Crimean mud doesn't wash out. My dress is ruined. Those Russians have a lot to answer for.'

Despite her bravado, Gideon was aware that the hand holding the match shook when she held it up to light her cheroot. Her close encounter with death had shaken her up, but Gideon doubted very much whether it would stop her from repeating the experience should such a need occur again. He thought she was probably the bravest woman he had ever met.

'Come along to the artillery observation post with me, Nell. We'll see if someone there can produce a cup of tea for you.'

'I'll see you there in a few minutes,' Nell replied. 'I want to see my patient on his way to hospital first.'

Turning to the sergeant and the soldiers who had come to her rescue, she said, 'There'll be a slap-up meal waiting for you at Nell's Nest when you can get there – but be sure there's someone to see you safely back to camp afterwards. You won't be leaving sober.'

19

I

'I've had a letter from Verity's grandfather today.'

Alice gave the news to Gideon when he came from the navvies' camp to Nell's Nest to have his evening meal with her at the end of another day's work.

The main railway link to the siege lines around Sebastopol was completed and the navvies were now working on the short, single-track tributary lines. These would be in operation within a fortnight and then Gideon's work in the Crimea would be done. Alice was aware that the time was fast approaching when she would need to make a decision about her own future.

'He must be distraught at losing both his son and his daughter-in-law, but what does he have to say about Verity?' Gideon seated himself at the table in the eating-house, where Alice had been waiting.

'He says he can't have her.' Gideon could see now that Alice was very upset. 'It's not just Gwen and John he's lost. His wife died in the same week they did. I feel so sorry for him. Jack was his only son and it seems there are no other close relatives. Mister Dymond is so upset he's thinking of selling the farm and moving into a small cottage, although I think he might change his mind when he's got over the shock of all that's happened. However, he's willing to pay the cost of having someone take care of Verity and says he will try to see her as often as he can, but it will be impossible for her to live with him.' Alice looked at Gideon despairingly, 'What am I going to do, Gideon?'

'What are *we* going to do,' Gideon corrected her. 'I've grown very fond of Verity too. Well, at least her grandfather doesn't intend disowning her – but he doesn't know her very well, and with so many things happening about him it must be difficult for him to make any plans right now.'

'True, but he's said nothing about arranging for her to go back to England – and I have no intention of sending her all that way just to be put to live in a house with complete strangers. She might be taken in for no other reason than to get money for keeping her.'

'*You* could offer to take care of her, Alice.'

'That's all right now, while I'm at Nell's Nest, but this war won't last for ever. How can I look after her back in England? Her grandfather is responsible for her and although he might be willing to pay for *her* keep, he's not going to support me as well.'

'Do you really need to ask such a question, Alice?' Gideon was hurt by her words. 'I know you haven't given me a definite answer yet, but I believed we had an understanding that we'd marry once we got to know each other better. I doubt if there are many couples who go into marriage knowing more about each other than we do. We've seen each other at our best and worst, in sickness and health, and in situations that most couples couldn't even imagine, and I'm more eager to marry you than ever. If we were married we could give Verity a far more secure home than either you or I enjoyed – and make sure she grows up knowing her grandfather, and he her. Of course, if you're still not certain about marrying me, it's a very different matter.'

'What sort of foolish talk is that, Gideon? You must know how I feel about you, and always have, for that matter.'

Gideon was overjoyed by her declaration, but he said, 'Yet you still haven't said you'll marry me.'

'Not in so many words I haven't, but there's a very simple explanation for that. When did you last *ask* me to marry you?'

'Ask you? You've always known how I feel about you.'

'I *think* I've always known, but I repeat my question. When did you last ask me to marry you?'

'That's a silly question, Alice. Why . . . I . . . don't know,' he admitted lamely.

'Exactly!' Alice said. 'From the very first day you arrived in the Crimea you've just *assumed* that we would be married one day – but you haven't actually *asked* me to marry you.'

Gideon was about to dispute her statement, but when he thought about it he realised she was right. He had suggested marriage when they were in England, and repeated his belief that they *should* marry in his letters to her, but he had never actually *asked* her to marry him. Certainly not since they had been reunited in the Crimea.

'You're probably right, Alice, but I haven't had much experience of this sort of thing.'

'I'm very pleased to hear it!' Alice retorted. 'I hope you're equally pleased to know that I'm just as inexperienced in the matter of proposals.'

In the silence that followed her statement, Alice looked at him expectantly. But Gideon found himself strangely tongue-tied. Alice waited for what seemed to her an age. Then tears welled up in her eyes and she stood up to leave.

'Alice!'

She hesitated and he stood up to take her arm and turn her towards him.

'Alice . . . will you marry me . . . please?'

'You don't have to . . . not if you don't want to.'

Confused by her sudden apparent mood swing, Gideon was uncertain what he was expected to say. 'Alice, you just said . . . I . . .'

'Only if it's what you really want.'

She was still tearful, but at least she was not trying to pull away from him and both were oblivious of the attention of every one of the army officers in the restaurant.

'Alice, I've wanted to marry you since I first saw you in Cornwall. I asked you then, and I've repeated it in my letters. I've just asked you again to marry me. What do I have to do to—'

'Yes.'

Her interruption brought his exasperated plea to a halt. 'Yes?'

'That's right. Yes, I'll marry you.'

Now it was Gideon's turn to question Alice's decision. 'Alice . . . you're not just saying that for Verity's sake?'

'Do you really believe I'd do such a thing, Gideon?'

Alice was close to tears yet again and Gideon said hurriedly, 'No, I don't.' He almost ruined the denial by adding, 'Even if it *was* so it wouldn't matter. I love you and I want you to marry me more than I've ever wanted anything. More than I ever will want anything. But – you're quite sure? You won't change your mind?'

'Oh, Gideon, do I really need to answer that? You've just told me what I've been waiting to hear you say for such a long time. I love you too. Yes, I'll marry you. Yes, yes, yes!'

Then he was holding her close and the cheers of the officers who had been interested spectators of the proposal brought Nell hurrying from the kitchen. When she arrived in the restaurant Gideon and Alice still stood with their arms about each other, but both were embarrassed at the realisation that they had been the centre of so much attention.

'What have you two been up to?' Nell demanded.

'Alice has just agreed to marry me, Nell,' Gideon said happily.

'Is that what's the cause of all the excitement?' Nell exclaimed. 'I thought we must at least have taken Sebastopol.'

Alice reacted to her remarks uncertainly and Nell's face broke into a smile. 'Oh, come here, the pair of you. Let me give you a hug and a kiss.' Shouting over her shoulder, she called, 'Ali! Bring the champagne in here . . . *all* of it. We're going to have a party!'

II

Gideon would have liked to marry immediately, while he and Alice were still in the Crimea, but she had other plans – plans that he found particularly surprising in view of all that had happened to her before she left England.

Alice wanted to be married in the church at Treleggan.

She explained her reasons to Gideon a few days after their party at Nell's Nest.

They were once again enjoying an evening meal together, this time in a room in the private living quarters. The restaurant was crowded with French cavalry officers, celebrating the award of well-deserved gallantry decorations to some of their number. Nell had recruited a number of French army wives for the occasion and they were being assisted by wives of British soldiers and by the omnicompetent Ali, who helped everyone, including himself, in his role as wine waiter for the evening.

'I would have thought Treleggan is the very last place you would choose to begin married life,' commented Gideon, deeply disappointed that Alice would not marry him right away.

'It's the people who lived in Treleggan who made me unhappy,' Alice explained, 'not the church. Parson Markham loved the church and so did I, in my own way. I would often go in there when no one else was around. Sometimes it was to pray, at other times to gather my thoughts and get things into perspective when I was particularly unhappy. I certainly wouldn't have married there had Reverend Bushell still been the rector, but as well as the letter from Jeremiah Dymond I had one from Reverend Brimble, the rural dean. He said that Bushell and his wife left Treleggan suddenly, to take up a living in the West Indies, where it's warmer. He says it isn't intended to appoint another rector immediately, so Reverend Brimble is taking services there on two Sundays a month. I'd like him to marry us. He and Mrs Brimble were very kind to me when I first left Treleggan. I know that my marrying in Treleggan church would have made Parson Markham very happy too. I'd feel I was doing something very special for the man who was the closest I ever came to having a father.'

Aware she had deep feelings about it, Gideon said reluctantly, 'All right, Alice. If that's what you really want, it's what we'll do. But it means putting our wedding off for a long time – and it means being parted again, because the railway is coming to an end and I'll need to return to England once everything is cleared up here. There might be a new rector at Treleggan by the time we are together once more.'

'Not if I come back with you,' was the surprising reply. 'I've been thinking very seriously about it ever since you asked me to marry you. We could all travel home on the same boat. You, me and Verity.'

'What will Nell think about such an idea?' Gideon asked. 'She's come to rely on your help here.'

Alice smiled happily at him. 'I've already talked it over with her. She certainly needed me when we first worked together, but thanks to you and your navvies – and the improving weather – there isn't nearly as much nursing to be done now. Food and supplies are getting through to the soldiers on the siege lines and the warmer weather means far fewer are falling ill. Besides, now some of Miss Nightingale's nurses are working in Balaklava, the hospital patients there are being properly looked after.'

'But what about this place? Nell's Nest will be more popular than ever when it's easier for off-duty soldiers to get here.'

'You can see what it's like tonight, Gideon. There are dozens of soldier's wives eager to make a few extra shillings, either in the restaurant or working in the kitchen. Nell need never be short of helpers.'

Gideon felt very much heartened by her words. 'Well, if you're quite happy about leaving Nell and the Crimea we'll travel back to England as soon as I can get everything completed here. I suggest you write to Reverend Brimble and ask him about having the banns published for our wedding – and whatever else we need to do before we can marry. Why, I feel happier already. Perhaps we should have some more of that champagne Nell gave us the other night!'

Alice leaned across the table and kissed him. 'Thank you for being so understanding, Gideon. It really does mean a lot to me. By marrying in Treleggan it will almost seem that Parson Markham is there, giving me his blessing.'

Work on the Crimean railway was finally completed in April 1855. Its completion meant that the disaster which had threatened the very survival of the British army had been averted. Navvies, who had been a feared and despised class of men in their own country, were now being hailed as patriotic heroes.

Although Gideon's work in the Crimea was done, he was now faced with the task of arranging passage home for all the navvies. Sailor and the others of the gang he had led for so long would return to work in Cornwall, while Gideon remained behind to supervise the gathering together and shipping of any equipment not required by those who now operated the railway.

Because he no longer had the help of his navvies, this task took far longer than he had anticipated. Gideon was supervising the job alone, Ranald MacAllen having left some weeks before to put his talents to use in Canada, where the railway system was expanding at an exciting rate.

Gideon had heard nothing from Robert Petrie for some weeks and was becoming seriously concerned when, early in June, a telegraph message reached him from the contractor. Gideon was to return to England, where Petrie had negotiated a lucrative contract on his behalf to lay a second track alongside the single railway line already in place – in Cornwall. Petrie was now a Member of Parliament and intended taking on no new railway work for the present.

The contract suited Gideon well. As well as being recognition for the superb work he had done in the Crimea, it meant a return to Cornwall. Once there he could arrange to have an address somewhere in Treleggan parish, thus avoiding the complications that had arisen in respect of having the banns called for his wedding to Alice. Reverend Brimble had written to Alice to say he would be delighted to marry them, but it would be necessary to obtain a special licence from the bishop if neither she nor Gideon had an address in Treleggan.

Verity was too young to know what was going on, but she reduced Alice to tears when she expressed concern about going away in a ship, in case her mother might not be able to find her.

When the time came to leave, Nell's farewell to the departing trio was an emotional affair. However, although there was less demand for her medical skills right now, she prophesied that she would be sorely needed when the army finally made its assault on Sebastopol.

In the meantime, there was no doubt that the popularity of Nell's Nest would keep her fully occupied.

20

I

'I never thought I would be so thrilled just to watch ordinary people going about their everyday business!'

Alice made the comment to Gideon as they stood on the deck of the steamship *Cormorant* which was making slow progress up the River Thames, bound for Deptford. She was looking shorewards, to where traffic and pedestrians passed along a roadway close to the riverbank. Verity was riding piggyback on Gideon's shoulders, waving to a man and a boy who stood at the water's edge watching the ship pass by.

'They waved back at me!' a delighted Verity shouted excitedly and Gideon smiled up at her, at the same time saying to Alice, 'It's certainly a far cry from the Crimea and its horrors.'

'Yes,' Alice agreed, suddenly sad. 'Sometimes I wake up in the morning and think I must have been having a bad dream about it all – but then I see Verity and know it was only too real.'

Alice linked an arm through Gideon's and squeezed it to her affectionately. The voyage home had been very different from the outward passage in the company of the Sisters of Mercy and the other nurses. The weather had been hot and the sea reasonably calm for much of the time. They had broken their journey briefly at Scutari and there met up with Sister Frances who was delighted to learn they were returning to Cornwall to be married.

The sister had been saddened to learn of the fate of Gwen and her husband, but expressed great relief that Verity would not be going into an orphanage. 'She's such a charming little girl,' she said. 'It would be dreadful to think of her growing up not belonging to anyone.'

While in Scutari there was also a surprise meeting with William Turnbull. Seemingly forgetting all that had occurred in Balaklava, he greeted both Gideon and Alice as old friends with whom he had always been on the very best of terms. Shaking hands with Gideon, he was effusive in his praise for what he and his navvies had accomplished in the Crimea.

'It can rarely have fallen to the lot of one man to perform such valuable service for his country and his fellow men in their time of need. Indeed, had you not come to the rescue of the British army, our brave soldiers – and our country too – might well have suffered unthinkable humiliation. The country owes you its gratitude, my boy. I have told them so in my despatches to London.'

Turning to Alice, Turnbull beamed at her in a manner she would have found menacing had Gideon not been with her. 'As for you, my girl, I have held you up to the readers of my newspaper as an example of everything that is splendid in English womanhood. Your unstinting devotion to wounded British soldiers is an inspiration to every woman in the land. The fact that you have taken it upon yourself to care for poor Verity is proof of your compassion. It has not gone unnoticed, my dear. There is not a reader of our country's newspapers to whom the name of Alice Rowe is unfamiliar.'

William Turnbull's words had been dismissed by Alice and Gideon as the drunken ramblings of a man who was trying to impress upon them his own importance and they had quickly put him out of mind. The correspondent could not have been farther from their thoughts as the pilot of the *Cormorant* steered the ship confidently around the many bends of the Thames until, eventually, the wharf where the ship was to berth came into view.

The estimated date and time of arrival of the steamship had been telegraphed from Scutari and updated from Malta. In addition, their details had been taken as the ship passed by the signal station on the Lizard, in Cornwall, and telegraphed to London, so the *Cormorant* was certainly expected. What those on board had not anticipated was the band playing on the quayside as they berthed.

Gideon and Alice had everything packed ready to disembark and they took Verity on deck to see and listen to the band. They were in no great hurry to go ashore, but when the ship was safely moored Gideon was pleasantly surprised to see Robert Petrie making his way up the gangway to the ship.

Pointing him out to Alice, Gideon hurried to meet the new Member of Parliament as he stepped on board the ship. 'Mister Petrie! What a pleasant surprise!'

'It's good to see you too, Gideon. Congratulations on a splendid job. Well done, indeed. But where are Alice and Verity?'

'They're farther along the deck, listening to the band. Is it playing to welcome home the convalescing officers who travelled from Scutari with us? I've done no more than pass the time of day with them, but I presume they must be heroes to be given such a welcome.'

Petrie looked at Gideon strangely. 'The welcome isn't for any army officers, Gideon – it's for you and Alice. Verity too, but especially for Alice. Your friend William Turnbull has fired the imagination of the country with his reports of "the Angel of Balaklava". He's also run a great many stories about the remarkable work done by the navvies under your leadership – and of your part in the "Battle of Nell's Nest", as the newspapers called it. The battle that left Verity an orphan was also reported in great detail. As a result, when it was learned that you and Alice are to be married and were returning to England *with Verity*, it was decided you should all be given a welcome befitting heroes.'

Gideon paled. 'You mean . . . all this fuss is for us?'

'This "fuss", as you call it, is only the beginning, Gideon. You and Alice will be fêted by the whole of London society. Your arrival has been anticipated with a great deal of excitement. I wouldn't be surprised if you were invited to meet the queen at Buckingham Palace.'

'Oh my God!' Gideon was genuinely frightened. 'All Alice and I wanted was a quiet return home and a simple wedding in Cornwall, as soon as it could be arranged. We're neither of us prepared for all this!'

'Well, let's go and put Alice in the picture and see what she has to say about it. Then we'll go ashore and – quite literally – face the music. You, Alice and Verity will be staying at my London home while you're in the capital and my wife, Sonia, is on the quay to provide support for Alice and Verity. You'll be pleased to know that a number of the Crimean navvies have come here today to welcome you home too. I've no doubt Sailor would be here, but he's in Cornwall, busying himself on the contract I've negotiated on your behalf. But here's Alice. She's no doubt wondering what's keeping us talking for so long. Let's break the exciting news to her . . .'

II

Gideon and Alice had not intended spending time in London but the unexpected publicity given to them by William Turnbull and the London newspapers meant they were now celebrities. As a result, they were obliged to remain in the capital for a few hectic weeks and were fêted wherever they went.

Taken to the House of Commons by Robert Petrie, they received a standing ovation from those in the chamber and, constantly being interviewed separately and together by representatives of the press, they were required to attend a bewildering number of functions.

Photography was still in its infancy but it had already become very popular and photographs of Alice, Gideon and Verity were taken at every event they attended and distributed all over the country.

The immediate result was embarrassing when the now-famous trio went on a shopping expedition. The owners of the shops they visited refused to accept money for the goods they chose, even when Gideon took Alice to a high-class jeweller's to buy the engagement ring that had been unobtainable in the Crimea. Gideon felt this was one item he really should pay for but the jeweller would not hear of it. It was, he declared, 'an honour to know that one of his rings would adorn the finger of "the Angel of Balaklava"'.

The home of Robert and Sonia Petrie proved to be a haven during this time and the two women became firm friends. Sonia grew particularly fond of Verity. The former contractor's wife longed to have a child of her own, but she and Robert had so far been unsuccessful. She hinted that if any problems arose in bringing up Verity, she and her husband would seriously consider adopting her.

Alice realised that such an arrangement would guarantee Verity an assured and happy future, but she too loved the small girl – and there were Gwen's wishes for her daughter to be considered.

Such was the enthusiasm of London for the trio from the Crimea that it was not until August, more than two months after leaving Balaklava, that Gideon, Alice and Verity were able to board a train to leave London behind. Heading for the south-west of England, they would complete their journey into Cornwall by coach.

'It's a relief to be free of people for a while,' Alice said, as they settled back in the otherwise empty compartment of the train.

Gideon smiled at her. 'I agree wholeheartedly, but you must admit we've had a very exciting time in London.'

'That's true,' Alice admitted, 'but I didn't feel there was a moment of the day when we could just relax and be ourselves. All I want now is to get to Cornwall and have time to think about ourselves, and our wedding.'

'Are you still sure you want to marry me, now you're famous?' Gideon asked, only half teasing. Alice had been subjected to far more hero-worship than he, but she had appeared to take it in her stride. There had even been times when Gideon believed she might actually be enjoying the attention.

'Marrying you is what I've really wanted from the first time we met,' she said seriously, 'even though I wasn't fully aware of it at the time.' She looked at him searchingly. 'I haven't changed – have you?'

'Not a bit,' Gideon declared fervently. 'But although you haven't changed your mind about the way you feel about me – and I'm very relieved to know that – you *have* changed, Alice. In the year we've known each other you've changed out of all recognition. You're no longer Alice Rowe, maidservant to the rector of Treleggan. You're the *famous* Alice Rowe, "Angel of Balaklava", and I'm very proud of you. You've experienced war and tragedy at first hand, and learned to cope with fame and adulation too. You could have anything or anyone you wanted.'

'I *have* everything I want, Gideon. What's even more important to me, I'll soon be married to you. That's what I want from life. As for changing . . . I'd love you as much if you were an ordinary navvy, but you're not. You've changed in the year we've known each other, too. From being a ganger, you're now a contractor, friend of a Member of Parliament, and famous for what you achieved in the Crimea. I'm glad to have you proud of me, Gideon, it makes me feel warm inside to think of it, but I'm just as proud of you. I knew you were someone special when you arranged that wonderful day to celebrate my twenty-first birthday, but I didn't know then quite how special you are.'

'Why are you and Gideon kissing?' The unexpected question from Verity broke the long silence in the railway compartment.

'I thought you were asleep, young lady,' Alice replied. 'We're kissing because we're going to be married and people who are married kiss each other a lot.'

'*I'm* not married, but you and Gideon give *me* lots of kisses,' Verity said seriously.

'That's because you're a very special little girl and we love

289

you very much.' Gideon smiled at her. 'I'm sure your grandpa will love you very much, too.'

'Mummy used to tell me lots about Grandpa Dymond . . . will she be there?'

Alice had tried to tell Verity in the gentlest way she could that she would not see her mother again, but Verity became so upset when she remembered the terrifying dawn attack on the day the Cossacks took Gwen away, she had decided to leave a full explanation of what had happened to Gwen until the memories were less vivid in Verity's mind. That time had not yet arrived – as a nightmare only two nights before had made very clear.

'I don't think so, poppet . . . but look, there are some baby pigs in that field. Don't they look sweet?'

The pigs successfully distracted the small girl and Alice looked at Gideon with an expression of despair. He shook his head sadly. He was as unsure as Alice about the best way to make Verity aware of the loss of her mother, and the need to deliver such enlightenment hurt. He hoped that time was on their side and would soften the blow when it had to come, but there was still not a day when Verity did not ask about her mother.

III

The route chosen by Gideon to enter Cornwall was via Tavistock and Launceston. Robert Petrie had told him there was talk of a railway line's being built along this route and Gideon wanted to have an idea of the land in the area, to inform his bid for any contract that might be offered in the future.

From Launceston, the trio travelled by coach across the beautiful and sparsely populated Bodmin Moor, heading for the county town which gave its name to the moor. Here they intended calling on the Reverend Harold Brimble, to make arrangements for an early wedding in Treleggan parish church.

It was late in the evening when they arrived in the town and Gideon booked them into two rooms at the Royal Hotel, from where they sent a message to the home of the rural dean to inform him they were in Bodmin and would like to call on him to discuss their marriage.

The reply came by return with the same messenger. Late as it was, Harold Brimble declared that he and Beatrice would have been delighted to receive them that very evening. However, he understood they would be tired after such a long journey and so he and his wife looked forward to seeing them the next morning, at whatever time was most convenient to them.

Beatrice Brimble's welcome was as warm as the woman herself. She hugged Alice and kissed Gideon and Verity, her expression one of genuine pleasure. Harold Brimble too seemed genuinely pleased to see them, informing them that he and Beatrice had followed their progress in the Crimea with great interest, through the columns of the local paper.

They were all given an equally enthusiastic welcome from Digger. It was almost as though he recognised Alice. The small dog seemed very happy in his new home, though it was quite evident he was being overfed. After being introduced to Verity, the dog was allowed to go out in the garden with her, Beatrice calling for a maid to accompany them outside.

'The local newspapers gleaned every word that was written about you from the reports reaching London,' Parson Brimble said, beaming at them. 'Cornwall is proud of you both.'

'I doubt if such news reached Treleggan,' Alice said. 'I can't think of anyone there who ever read a newspaper – unless it was Henry Stanbeare.'

'They didn't need to have a newspaper,' Harold Brimble said with some satisfaction. 'I ensured they were kept acquainted with what you were doing by pinning up the articles and news items about you in the church porch, along with the parish notices.'

'What did Henry Stanbeare think of that?' Alice asked. 'I can't believe he approved.'

291

'Mister Stanbeare's approval or disapproval is no longer a matter of any consequence in the parish,' Parson Brimble replied. 'In view of his behaviour and, in particular, that of his son, I was obliged to ask him to resign from his post of church-warden.'

Alice was intrigued. 'Resign because of Billy? Why, what has he done?'

Her question was answered by Beatrice Brimble. 'I wanted to tell you in a letter, my dear, but Harold said you had far more important things to attend to and would not want to bother yourself with petty Treleggan gossip.'

'That might have been true at the time,' Alice commented, 'but I'm not in the Crimea now. What has Billy Stanbeare been up to?'

Beatrice Brimble glanced quickly at her husband, and he shrugged. 'Well, dear, you might as well know that there were . . . rumours about Billy Stanbeare and the two Indian girls who worked for Reverend Bushell and his wife. Unfortunately, it transpired that it was more than mere rumour. We all thought the girls had returned to India, but it seems they got no farther than London. Apparently there is a large Indian community there. Somehow, one of their religious leaders learned of the plight of these two unfortunate girls and, after talking to them, he took them along to a magistrate. As a result, a warrant was issued charging Billy Stanbeare with . . . well, with rape. When the case came to court, here in Cornwall, it created something of a sensation. However, Henry Stanbeare employed a good defence solicitor and Billy was found not guilty.'

Angrily, Alice said, 'Anyone who knew Billy would have believed the girls, not him. He's capable of anything – and lying straight-faced about it.'

'I'm sure you're right, Alice,' Beatrice Brimble said soothingly. 'Unfortunately, the jury did not know him as you did. However, in protesting his innocence, Billy Stanbeare sowed the seeds of his eventual downfall. He was acquitted because the jury believed him when he said they consented to what took place between them. As a result, when they both had their babies he could hardly contest the . . . the bastardy orders that were taken out, naming him as the father! But rather than face up to his responsibilities, Billy disappeared.'

'Disappeared? Where did he go?'

This time it was Harold Brimble who replied. 'Opinion in Treleggan is divided,' he said. 'It seems he told some of his friends he was going to America, and others, Australia. One thing is quite certain: he is no longer in this country and, from all I have heard since, we are well rid of him. It came out that Henry Stanbeare had tried to buy the girls off when Billy was on trial for what he did to them. I felt this was not the sort of conduct the Church expects from one of its churchwardens. I feared my attitude towards him might result in fewer Treleggan parishioners attending my Sunday services, but it has had quite the opposite effect. What is more, with Billy gone from the village and his father discredited, I sense quite a different atmosphere in Treleggan.'

'It's a pity it didn't happen many years ago,' Alice said bitterly. Then, putting thoughts of the unhappy past behind her, she said, 'But Gideon and I are here to talk of happier things. We would like to make the arrangements for our wedding.'

21

I

It was two days after their visit to the Reverend Brimble that Gideon, Alice and Verity set off in a hired pony-cart to visit Jeremiah Dymond at Dewey Farm, on the extreme northern boundary of the Treleggan parish. It was not the most accessible farmhouse on the moor, and for the last part of the journey Gideon let the pony pick its own way along a very rough track.

He had spent the previous day with Sailor and his gang, who were hard at work on the railway that would eventually link Cornwall with the English counties beyond the River Tamar. Satisfied that the navvies were working as well under Sailor's supervision as they had for him, Gideon was happy to devote a day or two to Verity and her future.

Despite the state of the approach track, the buildings of the centuries-old Dewey Farm were in excellent condition.

'Verity's grandfather not only knows what he's doing, but he quite obviously enjoys farming,' Gideon commented, impressed with what he saw.

'That's what Parson Markham used to say,' said Alice. 'In fact, he said that Jeremiah Dymond was probably the best farmer in the whole of Treleggan parish. He liked him too. He often said that if he'd lived closer to Treleggan village he could have played a very important part in parish affairs.'

'Is this him?' Gideon asked quietly as the pony-cart came to a halt in the farmyard and a tall, grey-haired man emerged from the house to meet them.

'Yes . . . but he looks so much older!' Alice said as Jeremiah Dymond advanced to meet them.

Before Alice or Gideon could greet him, the farmer held out his arms and said, 'Verity! My very dear little girl! Come here . . .'

Jeremiah Dymond was a stranger to Verity and she looked to Alice for reassurance.

'It's your grandpa,' Alice whispered, 'and he's very happy to see you, poppet. Say hello to him.'

Verity allowed her grandfather to take her in his arms and hug her to him, but she was not entirely at ease, and Alice said apologetically, 'Verity's been through a great deal, Mister Dymond. She still isn't sure of anyone, or anything.'

'Of course not, I quite understand,' the farmer said. 'But she's such a lovely little girl, even lovelier than I imagined she would be. Her grandmother would have loved her so very, very much . . .'

His voice broke and Verity looked at him in sudden alarm.

Aware he had frightened her, Jeremiah Dymond forced a smile. 'It's all right, Verity. Your silly old grandpa is so pleased to see you that he's quite overcome. But come inside, all of you. I owe you a deep debt of gratitude. Especially you, Alice. Poor Parson Markham would have been so proud of you. He always said you were a cut above the village girls. From what I have heard of your exploits in the Crimea, he was most certainly correct. But come in . . .'

The inside of the farmhouse was tidy enough, but to Alice's expert eye it was apparent that it lacked a woman's touch.

Verity's grandfather had carried her inside the house and, when she asked to be put down, he released her reluctantly.

There was a cat asleep on one of the chairs and Verity hurried to it. Alice was concerned it might scratch her, but Jeremiah Dymond said, 'She'll be all right. Petal is a simpleton in the cat world, but there's not a penn'orth of harm in her.' While Verity cuddled the cat, he continued, 'I've heard much of your exploits in the Crimea. You're a very brave young woman.'

'Gwen was the brave one,' Alice replied. 'She was determined that nothing would stop her from joining her husband – your son – on the siege lines around Sebastopol.'

Jeremiah nodded. 'Unfortunately, she paid the price for such loyalty. I understand her body has never been found?'

'That's right,' Alice agreed unhappily. 'But a Cossack prisoner said that the bodies of a couple of the captured soldier's wives were thrown into a river and washed out to sea. It's probable Gwen was one of them.'

Jeremiah shook his head distressfully. 'Why do there have to be wars? Isn't there enough unhappiness in the world without them?'

In an attempt to change the subject, Alice said, 'I was very sorry to hear about the loss of your wife, Mister Dymond. Who is looking after you now?'

'No one,' came the reply. 'But it doesn't really matter. As far as I'm concerned the heart has gone out of the farm. I get by, although I intended giving up the farm at first. But it's been my life for so long I don't know what I'd do without it. It was to have been Jack's inheritance – his and Gwen's. Now I suppose that in due course it will pass to Verity. It's not exactly a rich estate, but if she were to marry a farming man it would give them and their family a living. I'd like to keep it going until then.'

As he was talking, a plan was forming in Alice's mind that might prove advantageous to everyone. 'Do you have any help in the house, Mister Dymond?'

'I don't need any,' he declared defensively. 'I can manage well enough, even if it does take an hour or two out of the working day.'

After casting a quick glance at Gideon, who realised where her questions were leading, Alice said, 'Mister Dymond, would it help if you had help for a couple of weeks in the house – and with the chickens and farmyard animals? Help that would not only give you time to get things on the farm into some sort of order, but also allow you and your granddaughter to get to know each other?'

'What are you suggesting?' Jeremiah looked at her suspiciously.

'I'll be absolutely honest with you, Mister Dymond. You'd be doing me and Gideon a very great favour. You see, we're getting married and I want it to be in Treleggan church, as a sort of tribute to Parson Markham. The rural dean, Parson Brimble from Bodmin, is quite happy to marry us there, but he says that in order for the banns to be called one of us, at least, must live in the parish. If I stayed here on the farm until my marriage it would give me an address in the parish, help you get things in order, and give you some time with Verity.'

Jeremiah Dymond had tried to present himself to those who knew him as a man who was coping with the deaths of his wife, son and daughter-in-law, but this was far from the truth. The loss of virtually the whole of his family in such a short period had hit him very hard. He spent so much of his time thinking about them that he was incapable of carrying out any task about the farm in the way it should be done.

He was an intelligent man and was aware of what was happening. Yet, try as he might, he could not stop mourning his lost family, even though it seriously affected his work. He realised that having Alice and Verity about the farm might be exactly what he needed in order to get to grips with life again.

But Jeremiah Dymond was a proud man. He was reluctant to admit he needed help.

'Well, if it means that you'll be able to marry in the way you want . . . I certainly owe both of you a great deal for what you've done – and are doing – for Verity, and I had a great deal of respect for Parson Markham. I've never been a great one for going to church, but that never stopped him from coming here to visit me whenever he could. He was a good man. All right, Alice, you're welcome to stay at Dewey Farm for as long as you need. You too, Gideon, whenever your work allows.'

Beaming at his granddaughter, he added, 'As for you, Verity, I have no doubt at all that you're going to brighten up the old place with your lively chatter.'

II

With the decision made to move to Dewey Farm the following day, it seemed to Alice the last barrier to her marriage to Gideon had been removed, and she was extremely happy.

As the pony-cart turned out of the farmyard with Verity waving and blowing kisses to her grandfather, Alice asked Gideon, 'Would it be possible to pay a call on Widow Hodge on the way back to Bodmin? I'd like to tell her our news and invite her to our wedding.'

Even as she spoke, Alice remembered that before they left London Robert Petrie had said he would be happy to come to Cornwall for their wedding, to act as Gideon's best man. She wondered whether Gideon would approve of having the disreputable old widow there too.

'. . . That's if you wouldn't mind, of course,' she added uncertainly.

'Mind? When I last saw Widow Hodge I promised her she would be the first to get an invitation!'

Gideon told Alice about calling at the moorland kiddleywink when he was on his way from Treleggan rectory to Bodmin in search of her. 'Widow Hodge has a soft spot for you,' he added. 'She was determined I should know what a bargain I'd be getting when I married you.'

'She's always been kind to me,' Alice said, 'though I don't know why she should be. She's never gone out of her way to be particularly nice to anyone else.'

Widow Hodge was inside the kiddleywink, but she saw them approaching and came to the door as they reached the house. Greeting them in typical fashion, she said, 'Well, I see you've finally managed to find each other, and now you've *both* come up in the world. The only other time anyone's come to visit me riding in a pony-cart was when one of them Customs officers called to find out whether I was selling smuggled liquor.'

Aware of the rumours that little, if any, of the drink sold in Tabitha Hodge's kiddleywink had ever passed through the hands of a Customs officer, Gideon asked, 'What happened, Widow Hodge? Were you able to satisfy him that duty had been paid on the drink you were selling?'

'The only thing I needed to satisfy was his thirst,' came the scornful reply. 'We sent him off unconscious in his pony-cart. I don't even know if he got back to where he came from – but he never troubled me again. Anyway, seeing as you're here, you'd better come inside and tell me how you came by that child you've got with you. I've heard lots of stories of what you've been up to in the Crimea, but I can't remember anything about a child. Come to that, I haven't heard about any wedding, either. Are you wed yet?'

'That's why we've come to see you, Widow Hodge,' Alice replied, 'to invite you to our wedding. We're having it in Treleggan church.'

'At Treleggan?' Widow Hodge looked back at Alice as she led them inside the house. 'I've no doubt it's a better place since Billy Stanbeare's left and Henry's got his come-uppance, but the mean-minded folk who should have got rid of the Stanbeares years ago are still in the village. I doubt if they're any better now than they were.'

'I don't care very much what the people of Treleggan think of me being married in their church,' Alice said defiantly. 'Gideon and me are getting married there as my tribute to Parson Markham.'

'The parson would have liked that,' Tabitha Hodge said, in unexpected approval. 'But now I suppose I'd better get you something to drink so we can celebrate the occasion and I'll see if I can find some fresh milk for this young lady – but only if she tells me what her name is.'

'I'm Verity.' Not entirely certain of the aged widow, Verity sought the reassurance of Alice's hand before adding, 'We've been to see my grandpa. We're going to live with him. He's got chickens, and pigs, and cows.'

'Has he now? Then he must be a rich man and there aren't too many of them about these parts . . . but let me fetch those drinks. Then you can tell me all your news.'

Gideon, Alice and Verity remained at the kiddleywink for close to two hours, exchanging news. The old woman was sympathetic with Verity's situation, but said, 'I know it's very sad, but she's luckier than most, Alice. She has you to take care of her.'

It was then that Alice asked the question that had been on her mind for so long. 'You've never had much time for the people of Treleggan, Widow Hodge, yet you've shown me nothing but kindness. Is there a special reason?'

'There is – and it's one them in Treleggan only believe they know when they put us down as being two of a kind. I was brought up in a workhouse too – and the same one as you, in Liskeard. I was with my ma and, like yours, she was always too sick to work. When she died I was nine and they gave me to a farmer to work for him as a dairymaid, but he wasn't a man like the parson who took you in. I was thirteen when I had his baby. It died within a fortnight because by then I'd been turned out by the farmer's wife and was living in the hedgerows, eating whatever I could steal from the fields. I buried the baby where it died, and afterwards couldn't even remember where it was, although I cried a lot whenever I thought about it. I moved about a great deal after that, working mainly on farms, but perhaps there was something about me that brought out the worst in men because they nearly all behaved like the farmer who first had me. But I'd learned something by then. I was using them too and I never went hungry again. Then I found my way here, to Bodmin Moor, and met William, my future husband. Life changed for the better then. He was no more of a saint than I was, but he treated me right and I never cheated on him. My only sorrow was that after all that had happened I was never able to give him the children he wanted.'

Aware of the effect her story was having on Alice, the old

widow gave her a mirthless smile. 'There, now you've heard a tale I've never told to anyone and I long ago outlived anyone who knew enough to tell it for me. Perhaps it's the reason I have a soft spot for miners and the like, and don't give a damn for farmers like Henry Stanbeare and his good-for-nothing son. It's also why I have sympathy for anyone whose early life was spent in a workhouse.'

Her story told, Tabitha Hodge shrugged off the past. 'But this isn't a day to be talking of such things. It must be that last lot of brandy I had brought to me. It's strong enough to loosen even a tongue like mine. Now, I know you're going to stay with Verity's grandpa for a few weeks, but you'll be wanting a home of your own. Do you have anywhere in mind?'

Gideon shook his head. 'Not yet, but we'd both like something not too far away from here. I hope to be working in Cornwall for the next year or two, but even if I get contracts farther afield I won't have to spend all my time on the workings. I've got some good men to look after things for me.'

'Well, if you're interested, I know of a place that might be just right for you. It belongs to Minnie Bolitho, a woman who's the closest to a friend I've ever made. Her husband was one of the men – an adventurer – who invested money in the mine where my William worked. We'd meet up sometimes, when they had a feast day or some other celebration, and although Minnie didn't seem to care too much for the other wives, she and I got on very well. Her husband owned shares in a number of mines and owned various properties too, but she'd sometimes come visiting here before it was a kiddleywink – and once or twice after, but by then Minnie's husband had died too and I'd go to see her at her home. It's only a few miles from here, over by Trenant. The house was a manor a couple of hundred years ago – Dowr Manor. It's certainly grander than a cottage, but it's not so big that you'd need an army of servants to take care of it. Anyway, Minnie sent a message to me a little while ago, asking me to go round and see her. She intends selling up and moving to London where she has a small house. I think she's forgotten how old I am. I can't go tramping all that way like I used to. But you go and see Minnie and tell her I sent you. She might sell the house to you if you like it.'

Alice was not at all certain that a house that had once been a manor was what she and Gideon had in mind and she looked at him questioningly.

Gideon was less inclined to dismiss Dowr Manor out of hand without seeing it first and he said, 'I have a better idea, Widow Hodge. I'll be moving Alice and Verity's things in to Dewey Farm tomorrow, but why don't I come here the day after with the pony-cart and we'll *all* go and look at Dowr Manor? That way you'll be able to see your friend before she leaves Cornwall.'

III

The day after Alice and Verity moved into the room they would share at Jeremiah Dymond's farm, Gideon called for them in the pony-cart. After collecting Widow Hodge they set off to visit Minnie Bolitho at Dowr Manor.

Tabitha thoroughly enjoyed the drive, as excited as a young girl at being taken on such an outing. The lanes around Dowr Manor were narrow and winding, with high banks, and they were almost upon the house before it came into view. Tabitha was the first to see it and called out excitedly, 'There it is. There's Dowr Manor.'

Alice gasped in disbelief and Gideon too caught his breath. The house was larger than Alice had imagined it would be, with tall chimneys which rose above the ridge of the roof. It was not built entirely of granite, which was the most common stone on and around the moor, but of a mixture of stones, including one which was slightly pink in colour, giving it a warm appearance, especially as much of the walls was covered in a broad-leafed creeper, the leaves already turning a deep red.

The gardens surrounding the house were a mass of colour from a wider variety of shrubs and flowers than Alice had ever seen growing together.

'It's *beautiful*!' she whispered, in awe. 'So very, very beautiful.'

'I said you'd like it,' Widow Hodge said smugly. 'You'll like the inside of the house too – but there's Minnie at the doorway, wondering who it is who's come visiting so unexpectedly.'

The woman standing in the doorway of the house appeared to be almost as old as Tabitha, and her sight was not good, but as they stepped down from the pony-cart and advanced towards her she cried out in delight.

'Tabitha! What a wonderful surprise, my dear. I never thought I'd see you again.' Minnie Bolitho hobbled to meet Tabitha and the two women hugged each other warmly. Then Minnie pulled away, but still clutching her friend's arm she asked, 'Who are these delightful people who have brought you to see me, Tabitha?'

'Oh, they are quite famous, Minnie – even young Verity. They've all just come back from the war in the Crimea. Alice nursed soldiers, and Gideon built a railway there.'

'Did you say *quite* famous? My dear Tabitha, they are very famous, very famous indeed. Even *I* have heard of Alice! As a matter of fact, I have a great-nephew with the army in the Crimea. He wrote to tell me of the Cornish girl who captured the hearts of his soldiers and my niece has sent me cuttings from the London papers which also mentioned the railway. My dears, I am honoured. Come inside the house. I am afraid you'll find it in something of a muddle. I've been packing up the things I will take with me to London as soon as I find a buyer for the house. I have had one or two offers, but I didn't like the people who wanted to buy. They talked of changing *this*, knocking down *that*, and building on all sorts of things. My niece thinks I am quite foolish and should accept whatever offer comes along, but she has never been one to look beyond practical considerations. This has been my home for more than sixty years. I have no intention of selling it to anyone who will make the house unhappy. I know everyone probably thinks I am just a silly old lady, but houses *do* know whether people love them or not, and they react accordingly. This has always been a happy home. The house and I love each other. I want it to be that way with its new owners.'

Tabitha exchanged glances with Alice and Gideon before saying to her friend, 'Do you know, Minnie, I might just have found the new owners you are looking for . . .'

The interior of Dowr Manor was as delightful as the exterior had promised. While Minnie Bolitho was talking, Alice's searching glance had taken in the huge open fireplace in the sitting-room, the centuries-old oak-beamed ceiling, the diamond-paned windows and all the many other features with which she had already fallen in love.

Minnie was asking Gideon a great many questions about his plans for the future, but she was also watching Alice, whose delight with the house was quite evident. Eventually, she said, 'Why don't you both take Verity for a walk round the house and gardens while Tabitha and I have a chat? We have a great deal of news to exchange. Take your time – and feel free to look wherever you wish. When you return we will have tea and some cake.'

As Tabitha had said, the house was not huge, but neither was it small. There were seven rooms on the first floor and a bewildering number at ground level, in addition to a large and well-equipped kitchen. There was also small but comfortable attic accommodation for three servants, which reminded Alice of the room she had occupied at Treleggan rectory.

There were no servants in the house at the moment and Minnie explained later that she had found other employment for the maids who had once worked for her. Until she moved, the daily cleaning of the house was being carried out by the wife and one of the many daughters of a farm labourer who lived nearby.

The gardens of the small manor house were evidence of Minnie's love of flowers, although there was also a well-stocked vegetable garden at the rear.

As they walked back to the house, Gideon asked, 'What do you think of it, Alice?'

'What can I say?' she replied. 'It's wonderful, Gideon, it really is.'

'I like it too,' Verity said positively. 'Are we going to live here?'

'I doubt it,' Gideon replied. To Alice, he explained, 'It would be impossible not to fall in love with it . . . but it *is* larger than I imagined it would be. I fear it will be far beyond what I can afford. I'm sorry.'

Aware of his disappointment, Alice linked her arm through his and said, 'I realised that early on, Gideon, but it's lovely to have been able to look round it. Perhaps we'll be able to find something similar, but smaller.'

When they re-entered the house, they found Tabitha and Minnie already drinking tea.

'There's plenty left in the teapot,' Minnie said, 'and the kettle's boiling away on the hob in the kitchen – but be careful of the handle, it gets very hot.'

'I'll take the teapot to the kitchen and put some more water in it, while Alice pours some milk for Verity,' Gideon said.

When he had left the room, Minnie said, 'Well, what do you both think of Dowr Manor, Alice?'

'It's an absolutely lovely house and garden,' Alice said wistfully. 'I don't think I've ever seen anything quite so wonderful.'

'I like it too,' Verity said. 'I like the flowers.'

'And so do I,' Minnie said to her. 'You have very good taste, young lady.'

When Gideon returned to the room with the refilled teapot, Minnie said, 'It seems that Dowr has the approval of your two ladies, Gideon, but what do you think of it?'

'I don't think I've ever seen a house I liked more,' he replied.

Minnie was watching him closely as he spoke and now she said, 'I sense that you have a reservation. What is it?'

Hesitantly, and somewhat embarrassed, Gideon said, 'Well, thanks to the contract I'm working on right now and the work Mister Petrie is going to put my way, I have good prospects. And working as a ganger, with a good gang, I was able to save money over the years . . . but I don't think I have enough for a lovely house like this just yet.'

'If you like it enough, don't let that stand in your way.' Tabitha

305

broke in on the conversation. 'I have more money than I'll ever be able to spend in the years I have left to me. There's no one to benefit from it when I'm gone. I'll loan you the money you need.'

'What's all this talk of lending, and not having enough?' Minnie asked indignantly. 'I haven't yet told you what I am asking for Dowr.' When her statement had effectively silenced her listeners, Minnie said, 'Although I wouldn't accept a million pounds for Dowr from someone who would not be happy here, you can have it for . . .'

She mentioned a sum that brought a gasp from Gideon.

Misinterpreting his surprise, Minnie said defensively, 'Of course, that includes the furniture and the things that I haven't already packed up to take with me.'

'My surprise isn't because I think it's too much,' Gideon explained hurriedly. 'The price you are asking is ridiculously low. I could afford it without borrowing money from Widow Hodge – but you could get much more . . .'

'I have already told you, young man, money is a secondary consideration. I want to leave Dowr happy in the knowledge that its new owners will love it as much as do I. I think you and Alice will. I also believe that Dowr will love you too. So, if you want it, you only have to say yes, and Dowr is yours. I can move out whenever you wish.'

Hardly believing their incredible good fortune, Gideon looked across the room at Alice and her expression made any questions unnecessary.

'Yes, Mrs Bolitho, we'll take Dowr. And should you ever make a return visit to Cornwall, Alice, I and Dowr Manor will always welcome you for however long you choose to stay. Thank you. Thank you very much.'

It had always been Alice and Gideon's intention to have a quiet wedding, attended by only a very few guests, such as best man Robert Petrie and his wife Sonia; Ivy Deeble, Parson Markham's housekeeper, and her sister; Sailor; Beatrice Brimble, and Tabitha Hodge.

306

However, Alice and Gideon greatly underestimated their new-found fame and popularity. In the event, the whole of the gang of navvies for whom Gideon had been ganger insisted on attending the service, as did the Mother Superior of the Devonport Sisters of Mercy.

Verity made a winsome bridesmaid, and flanking Tabitha Hodge were miners from the Wheal Endeavour. Far more surprising was the presence of Captain Gilbert with his daughter and granddaughter, whom Gideon and Alice had last seen on the eventful excursion to Falmouth and Truro a year before. It seemed they had read all the newspaper articles about the couple and decided to attend the wedding and pay their respects.

When she had made up her mind to be married in the Treleggan church, Alice had been convinced the Treleggan villagers would boycott the ceremony, but she was wrong. They packed the remaining pews and even stood at the back of the church to watch the former rectory housemaid being married.

It was a simple but moving service, during which the Reverend Brimble praised the part Parson Markham had played in bringing up Alice and suggested he would have been very proud indeed had he lived to see what she had achieved.

After the ceremony, the invited guests made their way to the rectory, where a quiet celebration had been organised. By prior arrangement, the navvies adjourned to Widow Hodge's kiddley-wink where they would feel far less restrained in their celebrations and where Gideon had arranged for all the drinks consumed to be charged to him.

It was an arrangement that suited everyone. Had the navvies remained in the village they might well have later clashed with the few remaining friends of Billy Stanbeare.

Inside the rectory, the small, select gathering was celebrating happily when there was a sudden disturbance at the front door of the house.

Those inside heard raised voices, but before they had time to wonder about the cause, the owners of the voices approached the room in which the guests were celebrating.

A moment later the door was flung open and an agitated woman burst into the room.

Alice did not immediately recognise her. Then, when she did, she let out an incredulous cry of disbelief.

It was Gwen Dymond.

22

I

The dawn raid by Cossacks on the British army's artillery positions around Sebastopol took the gunners completely by surprise. It was not the first such incursion, but British and French infantry formed a thinly manned line across the Crimean peninsula specifically to guard against such attacks from the Russian army which was massed to the north.

On this occasion, the sentries had failed to spot the raiders. In the dead of night the Cossacks had crossed the river into the territory held by the French and British, without being detected.

In the grey pre-dawn, they had formed up unnoticed behind the British sector of the siege lines, and as soon as there was sufficient light to pinpoint their targets they fell upon the artillery gun emplacements and the nearby family compound.

It was little short of slaughter. The fierce Russian horsemen put to death almost everyone they found. Soldiers, women and children. However, a few of the younger women were taken up and carried off with the Cossacks when their swift and savage raid came to an end.

Their work done, they galloped north, to fight their way to

309

Russian-held territory through whatever opposition might have been roused against them.

Along the way, two of the women they had taken prisoner proved to be too much of a burden and they were killed and dumped from the horses of their captors.

When the horsemen were crossing the river that formed the natural barrier between the two armies, one of the remaining four women tried to escape. Whooping noisily, a number of the lance-bearing Cossacks made a game of allowing her to almost reach the bank before they used their lances to drive her back into deeper water. When the 'game' began to pall, one of the horsemen ran his lance through her and her body was left to float downstream, to the sea.

On the Cossacks' arrival at the far bank, a small squadron of British lancers appeared on the British-held side of the river. There were far too few of them to attempt to cross the river and attack the Cossacks and, in a brutal act of bravado, some of the Cossacks raped one of the remaining captive women in full view of the British cavalrymen, to the noisy approval of the other Russian horsemen. The noisy 'entertainment' came to an abrupt end when one of the British lancers produced a carbine. Firing across the river, he succeeded in wounding one of the Russians.

The angry Cossacks promptly slit the throat of the woman and tossed her body into the river, where it floated after that of her countrywoman as the savage Russian horsemen rode off.

Gwen was one of the two surviving captives and she had no illusions about her fate.

That night, her fears were realised. She was brutally raped by two of her captors. Even so, she was more fortunate than the other woman, whose sobbing and cries continued for most of the night, while Gwen's two captors seemed intent upon keeping her to themselves.

The following morning the Cossacks split into two groups, Gwen remaining with one and the other Englishwoman taken by the second. Gwen never saw her again.

Her own situation was precarious in the extreme. For her first meal she was offered a bowl of liquid and black bread which was so hard it was inedible. Turning to the bowl, Gwen found to her dismay that it contained a vile concoction of heavily salted water.

The Cossacks laughed at the face she pulled when she tasted it and she thought the food and drink had been given to her as some form of cruel joke, but then she saw it was what the men around her were breakfasting on.

Soaking bread in the salt water she managed to swallow some of it, even though it tasted vile, but her tribulations were only just beginning.

The previous day she had been carried on the horse of one or other of the Cossacks, but today her hands were tied behind her and a long cord was fastened around her waist, the other end looped about the arm of one of her captors. When they set off, she was forced to walk, or occasionally to run, behind his horse.

Twice she fell over, suffering grazed arms and legs, and was forced to endure the jeers of the Cossacks, some of whom prodded her with their lances as she lay upon the ground. Gwen would have happily remained on the ground and allowed the Cossacks to kill her – but the thought of Verity brought her to her feet on each occasion.

Gwen realised that her husband must have been killed, along with most of the other gunners, but she was convinced Verity would have survived. Gwen was determined she would not be left an orphan if there was any way of preventing it – no matter what humiliations she herself needed to suffer in pursuit of that aim.

Her days with the Cossacks became weeks, and the weeks turned into more than a month as the fierce horsemen moved from one part of the Crimea to another. Gwen lost all sense of time.

It was not cavalry weather now and the horsemen became restive. Gwen quickly learned a few words of the Cossack language. The words for 'hurry', 'quiet' and 'we make love' – although there was no love involved in the act.

She also understood that certain words meant she was about to take a beating, although she never understood what the words were, in English, or why she was being beaten. She suspected it was simply because they were bored by the lack of action.

Unlike the British, there were no women accompanying the Russian horsemen, even though they were waging war on their own soil. Gwen felt the lack of feminine company deeply, especially when, two months after her capture, as far as she could ascertain, having completely lost all track of time, she realised she was pregnant.

It was another month before her captors became aware of her condition. It happened when her morning sickness attracted the attention of one of the older Cossacks who had experience with pregnant women.

His diagnosis created great merriment among the crude Cossack soldiers, and provoked fury in the more violent of her two captors. The beating he gave her was so severe that she suffered a miscarriage. Nevertheless, although desperately ill, she was still expected to cook and wash for the savage horsemen.

Had the Cossacks been on the move, as they so often were, Gwen would not have survived. But her grim determination, together with a ten-day halt due to bad weather, enabled her to make a recovery.

The fact that she had been pregnant and had the unborn life beaten from her seemed to have had a strange effect upon the less violent of her two captors. He still treated her as a harsh master might deal with a particularly stupid servant girl, but he no longer went out of his way to be brutal and would occasionally bring her a titbit of food, which the Cossacks had a mysterious knack of acquiring. For this reason, Gwen became more kindly disposed towards him. Unfortunately, when her other captor became aware of it, he was angry with his companion – and with Gwen.

The ill-feeling between the two men came to a head after the Cossacks were replaced by another regiment and moved deeper into Russia, away from the battle areas.

They were covering considerable distances each day and the

more benevolent of Gwen's two captors had acquired a wiry little pony for her. It was a good-tempered little animal and, for the first time since her capture, Gwen allowed herself to become fond of something.

The more brutal of her two captors was deeply resentful of the present given to her by his colleague and the two men began to argue violently about the pony. The argument raged for days and, although both men slept with her, the brutal man became increasingly violent in his lovemaking and she realised he was intensely jealous of his comrade.

In truth, Gwen hated them both, but she welcomed the change in attitude of the Cossack who had given her the pony.

II

In the interior of Russia, the Cossacks arrived at a town of some considerable size and were ordered to camp in fields outside the perimeter. It was here that matters came to a head between Gwen's two captors.

So close to a town, it was not long before cheap vodka began to circulate among the Cossacks. It had happened before and such nights as this terrified Gwen. Drink tended to bring out the very worst in the ill-disciplined Russian horsemen. She feared that if things got too out of hand, the men would turn on her and her two captors would be unable to prevent them from using her as they themselves had since she was taken prisoner.

Tonight, she had cooked a meal for her two captors and was preparing to serve it to the more considerate of them when the other returned to the cooking-fire, having quite obviously been drinking heavily.

Uttering the words she had come to dread, he demanded that she leave the cooking, because he wanted to make love to her.

The Cossack who was waiting for his meal protested and the two men began arguing. Although Gwen did not understand their language, it was evident that the drunken captor was accusing his companion of trying to get Gwen to himself, pointing out the things he had done for her. The pony he had acquired figured largely in his argument and it seemed he eventually threatened he would cut the pony's throat.

By now the argument had attracted the attention of other Cossacks, who were loudly voicing their individual opinions. The most drunken of them urged the violent captor to carry out his threat. Others, more sober, were trying to calm him.

Suddenly, the drunken Cossack stomped off angrily and returned carrying a sword – one he had gained in a skirmish with British hussars, earlier in the campaign. His intention was clear. He was going to cut the throat of Gwen's pony.

Gwen screamed and ran to the pony, meaning to untie the rope about its neck and allow it to run free and escape from the drunken Cossack.

The more sober of her two captors put himself between his companion and the pony and, when the drunken Cossack tried to force his way past him, there was a scuffle.

As though something had suddenly snapped in his mind, the drunken Cossack let out a cry of rage and struck his companion with the sword.

The wounded man staggered backwards with the drunken swordsman after him, and another blow was struck. This time it was so hard that the sword passed right through the other man's body.

Gwen screamed again and now others came running to restrain the drunken Cossack as he tried unsuccessfully to withdraw the blade.

The incident caused uproar in the camp. At its height there was a sudden disturbance at the edge of the camp and the noise subsided suddenly as a party of armed men came into the square formed by the blazing campfires.

There were a dozen or so of them, dressed in smart green uniforms and forming an escort for a man wearing so much gold braid that Gwen realised he must be a very senior officer – the first she had seen since her capture.

The officer took in the scene immediately and barked a question to one of the Cossacks. He received a shrugged and mumbled reply that must have been as unintelligible to the officer as it was to Gwen. Stepping forward, he struck the Cossack a hard blow with the flat of his hand on one cheek, then repeated it with a back-handed blow to the other.

This act produced a low rumble of protest from the watching Cossacks and the officer's escort immediately turned outwards and held their muskets hip high, pointing at the grumbling horsemen.

Crossing to where the dead Cossack lay on the ground, his body still transfixed by the sword, the officer used his high-booted foot to turn the man's head until his glazed eyes were staring lifelessly up at him.

Shifting his glance to the sullen Cossacks standing around him, the officer suddenly saw Gwen. He stiffened and, striding to where she stood, he spoke to her angrily. When she made no reply, he repeated the words even more loudly.

'I'm sorry, I don't speak Russian,' she said, trying to hide her fear of what this obviously important man might do to her.

The officer's eyes opened wide, and when he recovered from his astonishment he said, in halting but excellent English, 'You . . . are British?'

Now it was Gwen's turn to show surprise – and delight. 'Yes . . . yes, I'm English.'

The officer frowned. 'What is an Englishwoman doing with this . . . this rabble?'

'I was taken prisoner when they attacked the siege lines around Sebastopol.'

'*Sebastopol!*' The Russian officer looked at her in disbelief. 'But these men left the Crimea many weeks ago. How long have you been their captive?'

'I don't know. Three, four . . . five months. I've lost all sense of time.' Suddenly and unexpectedly, tears flooded Gwen's eyes as she said, 'I left my little girl behind there . . . my husband too, although he was probably killed in the attack.'

The officer's eyebrows were drawn together angrily. 'You

have been a captive for *months*? Where is your captor?'

Cuffing the tears from her eyes, Gwen replied, 'The man who is dead was one. The other is the one who killed him.'

'Were they fighting over you?'

'I don't know,' Gwen said. 'I think they were fighting because the one who killed the other was drunk.'

The officer's anger increased visibly. Turning to the assembled Cossacks, he berated them loudly, then said something to them that caused them to suddenly look elsewhere. In fact, he had asked them to produce the man who had killed his fellow Cossack.

One of the escort spoke to the officer and pointed out Gwen's surviving captor, who was still being loosely held by two of his companions.

The officer barked out an order for the Cossacks to bring the man to him. When they were slow in obeying, two of the escort moved forward, took the man from his friends and dragged him back to the officer.

Turning to Gwen, the officer asked, 'Is this the man who took you prisoner and killed his fellow Cossack?'

Gwen nodded.

'Does he mean anything to you?'

It was an unexpected question, but Gwen had no need to think about her reply. 'I hate him.'

Turning back to her drunken captor, the officer spoke harshly to him. When he failed to reply, the officer struck him as he had the other Cossack, earlier.

Her captor made a drunken comment that was clearly insolent, and at an order from the officer the soldiers who had taken him from the crowd forced him to his knees. Once again the officer spoke to the Cossack, evidently asking a question. The kneeling man looked up at him, but said nothing.

Angrily, the officer drew a pistol from the holster at his belt, held it to the kneeling man's head and repeated his question.

This time, the Cossack replied, saying, 'Da!' – 'Yes.'

Before Gwen or anyone else realised his intention, the officer pulled the trigger of his pistol. There was a loud report and

Gwen gasped in shock as her late captor pitched forward to the ground.

The officer now called something to the Cossacks assembled around them. In reply, some grumbled, but most merely shook their heads.

Contemptuously, the officer turned away and spoke to Gwen once more. 'Do you have any property in the camp?'

Gwen shook her head. 'No, nothing.'

'Then we can leave immediately.'

'Where are you taking me?' Gwen asked the question even though she did not care where she went. Nothing could be worse than all she had endured among the Cossacks.

'I have a house in the town. First, you can bath and we will find clothes and have you examined by a doctor. Then, as soon as it can be arranged, you will be returned to your own people.'

'You'll let me go back to the Crimea? To find my little girl?' Gwen pleaded.

The officer shook his head. 'I cannot promise that. All I can do is arrange a safe passage for you to the British embassy in Vienna, in Austria. The ambassador was a friend in happier days.' Feeling this perhaps required further explanation, he added, 'My father was ambassador to your country. I lived in London for some years and made many friends.'

Less interested in the story of her rescuer's life than in being reunited with Verity, Gwen said, 'But my daughter is in the Crimea. I want to go back there and find her.'

'Your daughter *was* in the Crimea,' the Russian said. 'You said you were captured some months ago and believe your husband is almost certainly dead. What would have happened to your daughter had she been left alive?'

His words made sense to Gwen, even in her present distraught state. 'She would have been sent back to England – to an orphanage.'

Had Gwen remembered the wooden medallion she had placed round Verity's neck, she would have had little confidence of its being found by anyone who would act on its message, even had they been able to read. England was the obvious place for her to begin her search.

'You are right, sir, and I owe you my undying gratitude – if not my life. The Cossacks would have blamed me for causing the death of one of their own and the man you shot would have killed me. He came close to it on more than one occasion.'

The officer looked at her pityingly. 'You are a remarkable woman. Captives of the Cossacks rarely survive for longer than a few days. They are a savage and lawless race. We must ensure that such tenacity for life does not go unrewarded.'

III

The officer who had rescued Gwen from the Cossacks was no ordinary soldier. Count Michael Malakhov was a member of one of the minor branches of the extensive Russian royal family and commander of the region in which the Cossacks who had held Gwen prisoner were camped.

His power was enormous and he used it to the full in helping Gwen. The doctor who examined her decided, in spite of her protests, that she would not be fit to travel for at least two weeks. During this time she lived in an annexe of Count Michael's house, with servants to attend to her every need. She bathed, fed well, and acquired a wardrobe of clothes that were superior to any she had owned before, and friends of the Count came to marvel at the woman who had survived for so long as a prisoner of the brutal and unpredictable Cossack horsemen.

Gwen was supremely grateful to the generous Russian nobleman who asked nothing of her in return. However, now that survival itself was no longer of paramount importance, she was increasingly concerned about what might be happening to Verity. When it was suggested she might remain as the Count's guest for a while longer, she insisted that she be allowed to travel on to Vienna, and from there to England.

Aware of the reasons for her determination to return to her own people as quickly as possible, Count Michael acceded to her wishes. He wrote a letter to the British ambassador in Vienna, and provided a carriage for Gwen and an army escort as far as the Austrian border. He also gave her ample money to get her to Vienna, and beyond. When she thanked him for his great kindness, the Count replied graciously that it was trifling compensation for all she had suffered at the hands of his countrymen.

He also gave her another letter, secured with his personal seal, saying, 'I can give you no money that will be acceptable in the countries of Russia's enemies, but the war will not last for ever. When hostilities cease and our countries are once more at peace, present this letter to whoever takes over our embassy in London and he will ensure that you receive compensation more in keeping with your suffering.' Smiling at her, Count Michael added, 'Who can tell? If I survive the war it is possible I may be that ambassador!'

Gwen's travels across Europe, by road, rail and boat, took more than two months. Unfortunately, after such a promising start, the journey steadily deteriorated.

The British ambassador in Vienna was impressed that the count should have taken such an interest in her and did all he could to help her on her way. After listening to her harrowing story with sympathy, he supplied her with what he thought were ample funds for the next stage of her travels.

Unfortunately, serious summer storms meant that the journey took her longer than expected and she was forced to sell most of her expensive clothes in order to eat. As a result she was less presentable when she eventually arrived at the next embassy, and consequently received less help.

By the time she reached Paris she had been forced to sacrifice food and appearance in order to pay for transport by the cheapest means possible.

At the Paris embassy they expressed doubt about the credibility of her story. She would have shown them the letter with the seal of Count Malakhov attached to it, but, such were the suspicions of the embassy officials, she feared she might be arrested as a spy.

Eventually, she was given money, but it was no more than the minimum that the wife of a serving soldier might expect to receive.

Gwen initially fared little better in London when she appealed to the War Office for money and information about her husband and Verity. They referred her to the artillery barracks at Woolwich. Here she had better luck and made the heart-lifting discovery that not only was Verity safe and well, but she was in England, with Alice!

The discovery came about after an initially depressing interview with an orderly sergeant who treated her claim to entitlement with considerable scepticism. It seemed that the regimental records showed 'Gwen Dymond and child' struck off the list of women carried on the regiment's complement when they were disembarked at Malta. He was unable to find a record to show they had been officially reinstated despite Gwen's insistence that they *were* on the regiment's strength in the Crimea.

Gwen was getting nowhere with the intransigent sergeant when an artillery captain limped through the orderly room on his way to an inner office. Gwen recognised him immediately as one who had been with the regiment when they sailed to the Crimea.

'Captain Fawcett, sir!'

Turning round, the captain looked at her uncertainly for a moment or two before saying, 'Why, bless my soul! It's Mrs Dymond, isn't it? Bombardier Dymond's wife. What are you doing here?'

'It's a long story, sir. A very long story.'

Looking more closely at her, the artillery officer said, 'I think you had better come to my office and tell me all about it . . .'

Once in his office, Gwen told Captain Fawcett her story from the time she and Verity had been put off the troopship at Malta. When she reached the point where the Cossacks had raided the gun positions outside Sebastopol, the captain called for the orderly sergeant.

He entered the office clutching a sheaf of papers and, when the captain asked him to find the casualty list for the Cossack attack, said, 'I thought you would ask for that, sir. I have it here.'

The captain scanned the papers, and then, looking apologetically at Gwen, he said, 'I am afraid Bombardier Dymond is on the list, ma'am – and so are you.'

'I'm not surprised,' Gwen said, upset by confirmation of what she already believed, that her husband had died in the attack. 'I was carried off by the Cossacks. They kept me for some months, until a Russian officer had me set free. I've come overland through Europe – but I'm particularly concerned about my little girl. I had a friend out there, Alice Rowe, who promised she would take care of Verity if anything happened to Jack and me, but I doubt if she ever learned about the attack.'

'Alice Rowe? *Nurse* Alice Rowe? The girl who helped Mother Nell Harrup to take care of wounded soldiers?'

Gwen was startled. 'Yes . . . you've heard of her?'

'I actually met her on the harbour at Balaklava, when I was on my way home with this.' Captain Fawcett rested a hand momentarily on his leg. 'But Miss Rowe is back in England now, hailed as a heroine – and she has an orphaned girl with her.'

Unable to contain her excitement, Gwen cried, 'The little girl . . . is her name Verity?'

'I think that could well be the child's name, but we must be certain. Sergeant, go to the officers' mess and tell whoever is there at the moment that I want all the newspapers for the last couple of months.'

When the sergeant had hurried away, Captain Fawcett explained to Gwen, 'We keep all the old newspapers. It helps officers who return from abroad to catch up with what's been happening in their absence. Now, while we're waiting, I'll have an orderly bring us some tea and you can tell me of your time with the Cossacks. It must have been terrifying for you . . .'

By the time the sergeant returned bearing an armful of newspapers, Captain Fawcett had a very good idea of what Gwen

321

had suffered at the hands of her captors. It was not so much what she said as what she did not say that told him something of the true story.

With the bundle of newspapers on the desk in front of him, Captain Fawcett picked up and discarded two before saying, 'Ah! Here we are . . . Miss Rowe returned to England with a Mister Davey whom she plans to marry, and, yes, an orphaned girl named Verity, whose parents were both killed in a Cossack raid on the siege lines around Sebastopol . . .'

'Where are they now?' Gwen asked excitedly. 'Is Verity here, in London? Does it say?'

Tantalisingly, the answer was not found until a number of newspapers later when, after describing Alice's movements in London in great detail, a report said, 'The celebrated Alice Rowe, heroine of the Crimea, has left London for Cornwall, where she is to be married to Mister Gideon Davey. Upon their arrival, the orphaned Verity Dymond will be reunited with her grand-father—'

Gwen stood up abruptly, interrupting Captain Fawcett's reading from the newspaper, 'I must go to Cornwall, to Verity . . .'

Startled, Captain Fawcett said, 'Surely you will want to rest up for a day or two, my dear?'

'No. I want to go now. I want to find Verity . . . but I have no money. Can the regiment let me have some?'

'It certainly can – and I will make an immediate collection among the officers. You are an artillery wife, Mrs Dymond. One day we will ask you to write the story of your ordeal and escape from captivity. It will form part of the regiment's history. Now, I am aware how eager you are to leave for Cornwall, but there will be no train at this time of day to get you there. Allow me to arrange accommodation for you in Woolwich, and transport to the station in the morning. By then my brother officers will have shown their admiration for your bravery and fortitude in a practical fashion that will be yours before you leave.'

23

Gwen's dramatic arrival at Alice and Gideon's wedding break-fast caused a sensation.

She had reached her father-in-law's farm and been told what was happening, and that Verity was Alice's bridesmaid.

Overjoyed at seeing his daughter-in-law alive, and under-standing her desperation to be reunited with Verity without further delay, Jeremiah Dymond harnessed up his pony to a small cart and took her to Treleggan rectory, where the wedding reception was being held.

The reunion between mother and child was not as joyful as Gwen had spent so long imagining it would be. Her ordeal at the hands of the Cossacks had aged her and her sheer joy at seeing Verity again frightened the small girl and she clung to Alice.

Gwen was heartbroken, but Alice and the matronly Beatrice Brimble comforted her, assuring her that, given a few hours, and in a quieter environment, Verity would realise that the mother she had thought she had lost for ever had returned to her.

Gwen's sudden appearance would be the talk of everyone at the wedding reception for the whole of their lives, but it put an end to the planned festivities.

Gideon and Alice took Gwen and Verity back with them to Dowr Manor that evening, leaving Harold and Beatrice Brimble to return Widow Hodge to her moorland kiddleywink, while Robert Petrie and his wife rode back to their hotel in Liskeard.

That night, after Verity had been put to bed and Gideon had gone to her room to check that she was all right, Gwen, still desperately upset by Verity's failure to immediately recognise her, said bitterly, 'It's all gone wrong, hasn't it, Alice? Verity no longer knows me and I've ruined your wedding day. It would have been better had I not tried so hard to stay alive when the Cossacks had me – and, believe me, dying would have been much easier then than living.'

'That simply isn't true,' Alice declared. 'Having you suddenly appear at the reception, when we all believed you dead, was the best wedding present anyone could have given to me. As for Verity . . . she's had a very exciting day. Seeing you was just too much for her to take. She's never stopped asking about you, Gwen, and used to wake up having nightmares about the Cossacks, crying out for you. I think that in her own mind she had just begun to accept she would never see you again – and then, suddenly, here you are! Give her a while – a very little while – and she'll be the happiest little girl in the world to have you back. But right now, seeing you again will have brought all sorts of memories flooding back to her that she's been trying hard to forget. I remember that when Nell's Nest was raided and Gideon wounded, the sight of the Cossacks made her hysterical.'

'I knew nothing about that,' Gwen said. 'I didn't realise the Cossacks had attacked again. How badly was Gideon wounded . . . ?'

When Gideon returned to the room, he and the two women talked long into the night about all that had happened to them. Suddenly, Verity appeared in the doorway. She was clutching a doll that had been bought for her in London by Sonia Petrie.

Going to Gwen, hesitantly, she asked tremulously, 'Why did

you leave me when those nasty men came and were hurting everyone?'

Aware that this was a crucial moment, Gwen contained the overwhelming urge to hug her small daughter and said, as evenly as she could, 'I didn't leave you, darling. Those nasty men took me away and wouldn't let me come back to you, no matter how hard I tried.'

Verity thought about this for a while and then said, 'Did they hurt you?'

Try as she might, Gwen could not prevent the tears springing to her eyes as she said, 'Yes, darling, they hurt me very much, but the worst hurt of all was that they wouldn't let me come back to you.'

'Were you very unhappy?' Verity asked earnestly.

'I was *very* unhappy,' Gwen said, aching to sweep her serious little daughter up in her arms and cuddle her, but knowing the right moment had not yet arrived, 'very unhappy indeed.'

Still serious, Verity said, 'I'm sorry you were unhappy, but there are no nasty men here, so will you stay with me for always now?'

'Oh yes, for always and always,' Gwen said fervently.

Suddenly giving Gwen a shy smile, Verity said, 'Will you tuck me up in bed, like you used to?'

With tears streaming down her face, Gwen said brokenly, 'Tucking you up in bed will make all the unhappiness go away for ever, darling. Come here, I'll carry you upstairs.'

As mother and daughter went from the room, Gideon and Alice heard Verity say, 'If we go and live with Grandpa, he won't let any nasty men come and take you away again . . .'

When the door closed after them, Alice tried unsuccessfully not to sound emotional when she said, 'It's been a most unexpected wedding day, Gideon, but it couldn't have been a happier one. For Verity to have Gwen back is nothing short of a miracle. To have it happen at any time would have made it a special day. Coming, as it has, today, on the day of our wedding . . . I can't even try to describe how I feel – but I'm going to miss Verity.'

'I don't think you'll miss her for too long,' Gideon said, putting his arms about her. 'This is a very happy house that's just crying out for us to have children of our own.'

Kissing her as he held her close, he added, 'After all, why else would it have seven bedrooms?'